Lecture Notes in Mechanical Engineering

Lecture Notes in Mechanical Engineering (LNME) publishes the latest developments in Mechanical Engineering—quickly, informally and with high quality. Original research or contributions reported in proceedings and post-proceedings represents the core of LNME. Volumes published in LNME embrace all aspects, subfields and new challenges of mechanical engineering.

To submit a proposal or request further information, please contact the Springer Editor of your location:

Europe, USA, Africa: Leontina Di Cecco at Leontina.dicecco@springer.com
China: Ella Zhang at ella.zhang@cn.springernature.com
India, Rest of Asia, Australia, New Zealand: Swati Meherishi at swati.meherishi@springer.com

Topics in the series include:

- Engineering Design
- Machinery and Machine Elements
- Mechanical Structures and Stress Analysis
- Automotive Engineering
- Engine Technology
- Aerospace Technology and Astronautics
- Nanotechnology and Microengineering
- Control, Robotics, Mechatronics
- MEMS
- Theoretical and Applied Mechanics
- Dynamical Systems, Control
- Fluid Mechanics
- Engineering Thermodynamics, Heat and Mass Transfer
- Manufacturing Engineering and Smart Manufacturing
- Precision Engineering, Instrumentation, Measurement
- Materials Engineering
- Tribology and Surface Technology

Indexed by SCOPUS, EI Compendex, and INSPEC.

All books published in the series are evaluated by Web of Science for the Conference Proceedings Citation Index (CPCI).

To submit a proposal for a monograph, please check our Springer Tracts in Mechanical Engineering at https://link.springer.com/bookseries/11693.

Steven Y. Liang
Editor

Manufacturing Technologies

Selected Proceedings of the 9th International Conference on Manufacturing Technologies (ICMT) 2025

Springer

Editor
Steven Y. Liang
George W. Woodruff School of
Mechanical Engineering
Georgia Institute of Technology
Atlanta, GA, USA

ISSN 2195-4356 ISSN 2195-4364 (electronic)
Lecture Notes in Mechanical Engineering
ISBN 978-3-031-95454-2 ISBN 978-3-031-95455-9 (eBook)
https://doi.org/10.1007/978-3-031-95455-9

This Springer imprint is published by the registered company Springer Nature Switzerland AG
The registered company address is: Gewerbestrasse 11, 6330 Cham, Switzerland

Organization

Conference Chairs

Steven Y. Liang (Fellow of ASME and SME), Georgia Institute of Technology, USA
Ramesh K. Agarwal (Fellow of IEEE, AAAS, AIAA, APS, ASME, CSAA, SME and ASEE), Washington University in St. Louis, USA

Conference Co-chair

Ramulu Mamidala (Fellow of ASM, ASME, SEM and SME), University of Washington, USA

Program Committee Chairs

Jing Wang, University of South Florida, USA
Xiaohong Lu, Dalian University of Technology, China

Program Committee Co-chairs

Xu Chen, University of Washington, USA
Jin-Soo Park, Sangmyung University, Republic of Korea

Conference Support Committee Chair

Wei Huang, Georgia Institute of Technology, USA

International Technical Committees

Albert Lin, National Chiao-Tung University, Taiwan
A. Elaya Perumal, Anna University, India
Adirek Baisukhan, Chiang Mai University, Thailand
Ahmed Mohammed, University of Birmingham, UK
D. Ramasamy, Universiti Malaysia Pahang, Malaysia
Hatem Mrad, Université du Québec en Abitibi-Témiscamingue, Canada
Himadri Chattopadhyay, Jadavpur University, India
Jonghwan Lee, Kumoh National Institute of Technology, South Korea
K. Philip Varghese, Saint-Gobain Abrasives, USA
Luca Iuliano, Politecnico di Torino, Italy
Man Zhao, Shanghai University of Engineering Science, China
Marco Castellani, University of Birmingham, UK
Nithin Tom Mathew, BITS Pilani, India
Omar D. Mohammed, Prince Mohammad Bin Fahd University, Saudi Arabia
Pradeep Kumar, Indian Institute of Technology Roorkee, India
Prasanna Shakti Jena, Nalanda Institute of Technology, India
Prasanna Shakti Jena, Vardhaman College of Engineering, India
Reza Nosouhi, Islamic Azad University, Iran
Rishi Pahuja, University of Washington, USA
Sanjeev Das, National Institute of Technology Raipur, India
Seung Hwan Park, Chungnam National University, Republic of Korea
Sicong Chen, GlobalFoundries Inc., Singapore
Subhas Ganguly, National Institute of Technology Raipur, India
Sudip Kumar Sinha, National Institute of Technology, India
Sujit Singh, Woxsen University, India
Sumesh Narayan, The University of the South Pacific, Fiji
Tommi Mahlamäki, Tampere University, Finland
Vijay Sekar K. S., Sri Sivasubramaniya Nadar College of Engineering, India
W. Caesarendra, Universiti Brunei Darussalam, Brunei
Wasawat Nakkiew, Chiang Mai University, Thailand
Xia Ji, Donghua University, China

Preface

Welcome to the Proceedings of the 2025 9th International Conference on Manufacturing Technologies (ICMT 2025).

It is our great pleasure to present the proceedings of 2025 9th International Conference on Manufacturing Technologies (ICMT 2025), held in Atlanta, USA, and simultaneously online during January 13–17, 2025. This year's conference brought together researchers, engineers, and industry experts from around the world to share cutting-edge advancements, innovative solutions, and future trends in Manufacturing Technologies and related fields.

ICMT has long served as a premier platform for fostering collaboration and knowledge exchange in Mechanical Principles and Engineering, Advanced Manufacturing and Industrial Automation, Material Design and Analysis and related themes. The ICMT 2025 continued this tradition, featuring keynote speeches by renowned scholars, technical sessions, and workshops, all aimed at addressing global challenges and driving technological progress.

This volume compiles a selection of high-quality peer-reviewed papers, covering diverse topics such as engineering material structure design and mechanical performance analysis, digital manufacturing system and control technology, and image-based intelligent control system and monitoring.

We would like to extend our deepest gratitude to The ICMT 2025 Conference Committees for their tireless efforts in planning and execution, to the Technical Program Committee and reviewers for their rigorous evaluation and constructive feedback, to the authors and participants for their invaluable contributions and vibrant discussions, and finally to our sponsors and partners for their generous support.

We hope this publication serves as a valuable resource for the field of manufacturing technologies and inspires further research and innovation.

Steven Y. Liang
Conference Chair
(Fellow of ASME and SME)
Georgia Institute of Technology
Atlanta, USA

Contents

Engineering Material Structure Design and Mechanical Performance Analysis

Enhancing Hole Characteristics During Magnetic Assistance Closed-Loop Electrochemical Discharge Drilling (M-CLECDD)

Viveksheel Rajput, Sarbjit Singh, and Perminderjit Singh

Abstract The significance of uniformity and surface quality is highlighted by recent advancements in the fabrication of holes on glass materials, particularly for industrial applications of the electrochemical discharge drilling (ECDD) process. Nevertheless, the problem of inadequate electrolyte in the hydrodynamic area prevents the process of drilling which further results into a high hole tapering and poor depth. The present research explores the application of magnetic-assisted electrochemical discharge drilling (M-ECDD) to produce holes on a glass material using a closed-loop system of adaptive tool feed method, implemented to prevent the tool contacts with the glass material. The comparative analysis of the hole taperness (HT), machining depth (MD), surface roughness (SR), and material removal rate (MRR) is carried out with and without magnetic effect. Moreover, the effect of various input parameters on hole characteristics is studied. It is observed that the magnetic field regulates bubble formation and detachment, improving the ECDD process. Lorentz forces affect bubble behavior and film stability, resulting in magnetohydrodynamic (MHD) convection which forces the electrolyte to travel at higher depths, improving electrolyte circulation and discharge frequency. An improvement of 14.86%, 58.14%, 32.71%, and 23.89% in MD, HT, SR, and MRR is observed with the application of the magnetic field.

Keywords ECDM · Magnetic · Spark · Convection · Hole tapering · Hole depth · Roughness

1 Introduction

Electrochemical discharge drilling (ECDD) is a precise method for producing miniature drilled components on a non-conductive material such as glass and ceramics [1]. High heat-affected zones are produced by heat-based methods like Electron

V. Rajput · S. Singh · P. Singh (✉)
Department of Mechanical Engineering, Punjab Engineering College, Chandigarh, India
e-mail: permindersingh@pec.edu.in

S. Y. Liang (ed.), *Manufacturing Technologies*, Lecture Notes in Mechanical Engineering, https://doi.org/10.1007/978-3-031-95455-9_1

Beam Machining (EBM) and Laser Beam Machining (LBM), which limits their use for a wide variety of industrial applications. For the purpose of machining non-conductive brittle materials that are difficult to machine with traditional methods, ECDD is a useful technology [2]. This hybrid technique may be used to turn, drill, and mill in addition to other operations by combining electrochemical and electro-discharge technologies. The process is known by various names such as electrochemical discharge machining (ECDM), die-sinking electrochemical discharge machining, wire electrochemical discharge machining (WECDM), electrochemical discharge trepanning, electrochemical discharge milling, and electrochemical discharge turning [3]. Because of its cheap cost, process flexibility, and generally better machining, ECDM has emerged as a viable option for drilling glass materials in light of recent advancements. However, obtaining higher hole depths with reduced tapering is still one of the challenges especially in hydrodynamic region [4]. ECDM, a hybrid between electrochemical machining (ECM) and electrical discharge machining (EDM), depends on electrochemical processes to start discharges that make chemical etching easier. The process consists of tool electrode (cathode) and the auxiliary electrode (anode), which are separated by a few centimeters inside an electrolytic cell as shown in Fig. 1. Using both electrodes, the glass substance is submerged in an alkaline electrolyte solution. A gas layer surrounds the tool as a result of hydrogen and oxygen bubbles forming and merging into bigger bubbles when electricity is applied. This layer separates the tool electrode from the electrolyte by obstructing the passage of electrons. Electrical breakdown through the gas layer causes sparks, which in turn cause material loss by thermal erosion and the electrolyte's etching action. Kurafuji and Suda [5] were the first authors to successfully demonstrate the use of electric sparks for material removal in 1968. Basak and Ghosh were the first researchers to investigate the importance of critical voltage readings to initiate machining and to explain the spark-generating process in the ECDM [6].

Several studies have investigated the effects of different input parameters such as tool's shape and materials, electrolytes and its concentration, applied voltage, and tool movement on discharge phenomena and subsequent machining performance in ECDD [7, 8]. Voltage and electrolyte concentration have a direct impact on MRR and surface roughness; higher values promote bubble formation and gas film production more quickly, which increases MRR and influences roughness [7]. Microchannel texture is strongly influenced by electrolyte viscosity [8]. To keep the machining gap constant, the tool feed rate needs to be regulated. Different methods such as gravity assisted and constant velocity are utilized to control tool feed [8]. Variations in electrical conductivity between different electrolytes impact MRR. To increase performance, ECDD has undergone a variety of modifications, including the use of tool rotation [9], abrasive particles mixed ECDD [10], magnetic field [3, 4], and ultrasonic vibration [8]. Every one of these modifications help to improve the machined materials' quality attributes. Recently, magnetic assistance ECDD has been implemented in various studies to improve the process. By causing magnetohydrodynamic convection (MHD) in the electrolytic cell, it integrates the magnetic flux with the ECDD process that further aids in the regulation of the formation of bubbles and

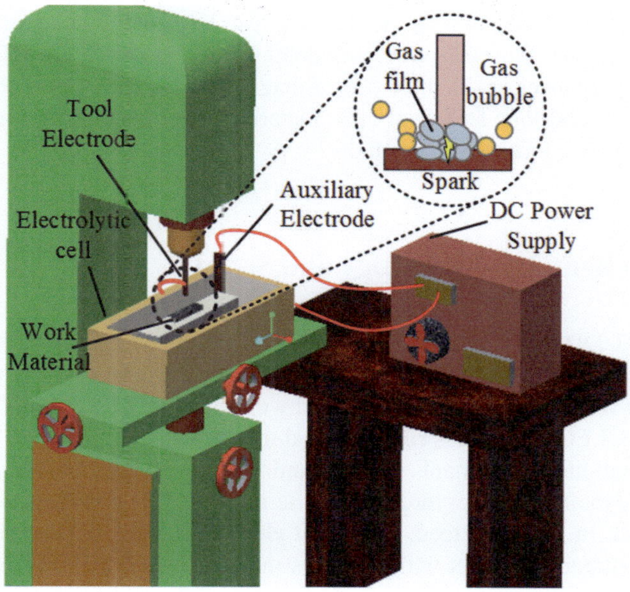

Fig. 1 ECDD working principle [7]

their detachments [4]. A common problem with deep drilling in ECDD is insufficient electrolyte circulation at the tool-work material gap. It is not possible to stop bubble nucleation and development using voltage, concentration, and other input variables alone. On the other hand, magnetic flux provides a steering effect that enhances the process through magnetic hydrodynamic convection [11]. The bubble is separated from the tool by steering action, which also reduces gas film thickness due to the magnetic field effect. Thus, M-ECDD process enhances the performance and quality of drilled hole using enhanced electrolyte circulation. A theoretical model describing the different physical processes involved in MHD convection was developed to examine the impact of MHD convection on microchannel fabrication [12]. Researchers produced microslits larger than 1300 μm using the MAECDM technique, and then used metaheuristic algorithms to find the best input parameters [13]. It is difficult to drill deeper than 1000 μm because to bubble accumulation. By reducing bubble formation and increasing electrolyte flow, MHD convection increases the accuracy of the machined surface [14]. For micro-drilling, the ECDM process's use of a magnetic field improves machining depth. High aspect ratio microchannels were created using multi-pass ECDM techniques [15]. Singh et al. [16] study found that a magnetic field enhances electrolyte circulation in ECDM, increasing machining depth and material removal rate, while decreasing taper angle and hole overcut.

According to the literature, various researchers have performed many experimental investigations on both the ECDM and M-ECDM processes. However, the investigations on HT, SR, and MD require more exploration especially when there is an issue related to the tool contacts with the work material in constant velocity tool feed method. Tool contacts with the material lead to the damage and deterioration of the hole features, especially at higher depths in terms of SR and HT. Therefore, the present study investigates the effect of magnetic assistance on HT, MD, SR and MRR during closed-loop ECDD of glass material based on novel adaptive tool feed method which has never been reported.

2 Experimental Methodology and Setup

The M-CLECDD technique is integrated and modified for the vertical milling machine as shown in Fig. 2. This setup consists of an electrolytic polycarbonate cell and non-reactive components that hold work materials. The study's developed and built M-CLECDD process runs in actual machining environments.

The limitations of constant velocity and gravity feed techniques have been addressed by the mechanism through the use of CLECDD process based on adaptive tool feed. In a gravity feed, decreased electrolyte availability resulted due to the tool's continuous contact with the glass material, which inhibits the formation of a gas layer under the electrode and resulted in the uneven material removal. Mismatched MRRs and feed rates might potentially harm the tool or material when moving at a constant velocity. Hajian et al. [11] discovered that the tool often comes into touch with the glass substance due to thermal expansion. In order to solve this problem, an adaptive system modifies the tool feed in accordance with the MRR in order to maintain an effective machining gap. It detects tool contact using a sensitive load cell and transmits signals to an Arduino UNO, which tells the controller to change the tool's position. Experiments are performed based on one-factor-at-a-time (OFAT) to analyze the effect of magnetic assistance on hole characteristics in terms of MD, HT, and MRR. KOH is used as an electrolyte for drilling a glass for 4 min. A total of nine drilling operations are carried out and repeated three times under various

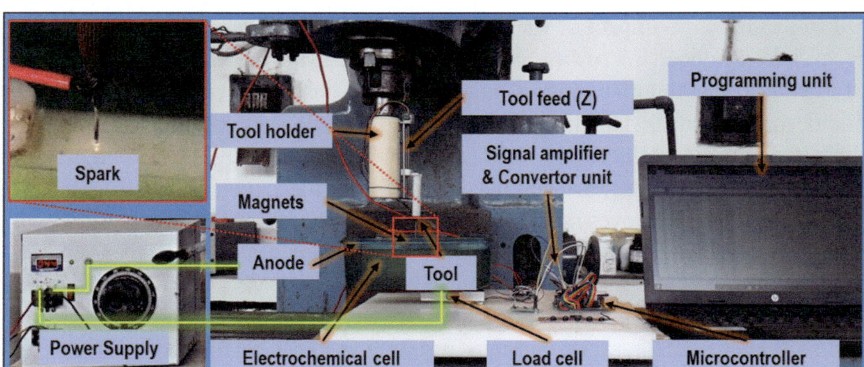

Fig. 2 Developed experimental setup of M-CLECDD process

Table 1 Drilling parameters and their levels employed during M-CLECDD process

Levels	Range	Fixed parameters	
Tool feed rate	1–4 mm/min	Work material	Soda lime glass
Applied voltage	35–55 V	Tool	Tungsten carbide (1000 μm diameter)
KOH concentration	10–30 wt%	Magnetic intensity	0.33 T

machining conditions. The drilling parameters that were utilized during M-CLECDD experimentations are listed in Table 1.

2.1 *Measurements*

After drilling holes on the soda lime glass material, the responses are measured to analyze the effect of magnetic assistance and effect of input parameters on hole characteristics. Measurements of MD, HT, and MRR are performed three times and the average value is utilized as the response value. Averaging the three measurements reduces computational errors, and the variability in the findings is within acceptable limits. Measurements of MRR are performed using a weighing machine (Model: CAY220, Make: CAS) with a precision of 0.0001 g. A ZEISS Stemi 508 stereo zoom microscope with software is used to analyze the MD and HT of the holes. High-to-low temperature tribometer is used to analyze the SR of the holes. MRR is computed by utilizing weight difference of soda lime glass before and after the M-CLECDD process in a unit time. The formula is: MRR $= ((w_{g1} - w_{g2})/t)$, where w_{g1} = weight before M-CLECDD and w_{g2} = weight after M-CLECDD process. A tapered hole is one in which the hole entrance diameter differs from the hole exit diameter. It is calculated by taking into account the hole entrance diameter (D_1), hole exit diameter (D_2), and hole depth or machining depth (h_d) as given in Eq. 1.

$$\text{Hole Tapering (HT)} = \left[\tan^{-1}\left(\frac{D_1 - D_2}{2 * h_d}\right)\right] * \pi/180\,\text{radian} \tag{1}$$

3 Results and Discussions

3.1 *Analysis on the Magnetic Field Effect on Hole Characteristics*

Microscope images are analyzed to compare the geometrical characteristics of the holes with and without the M-CLECDD process as shown in Fig. 3. The through holes drilled using the magnetic effect in ECDD at 35 V, 15 wt% KOH concentration, and a tool feed rate of 2 mm/min are circular, less tapered, and higher in depth than those drilled without the magnetic effect. A better hole shape is obtained with the help of magnetic assistance to the CLECDD process (Fig. 3d). The assistance of the magnetic field guarantees a steady supply of electrolytes and stable gas film development at higher depths which results in the uniform spark dispersion and reduced stray sparks. This improves the HT of the holes by 58.14% (Fig. 3a, d) and MD of the hole by 14.86% (Fig. 3).

The MHD effect, which is induced by magnetic flux, lowers the bubble departure radius and produces a thin layer surrounding the tool. This reduction in layer thickness and energy channelization reduces the discharge intensity, which is critical for developing deep holes. MHD convection influences film thickness, bubble rotation, and electrolyte flow, which improves chemical interactions and increases current density. As current density increases, the bubble separation size reduces resulting in increased discharge frequency. Subsequently, an improvement in hole taperness and depth is observed. The adaptive tool feed system also enhances the M-ECDD process by ensuring a constant machining gap. This guarantees that electrolytes are always present beneath the tool, which is essential for the consistent formation of the gas layer, resulting in more precise and accurate machining results. A decrease

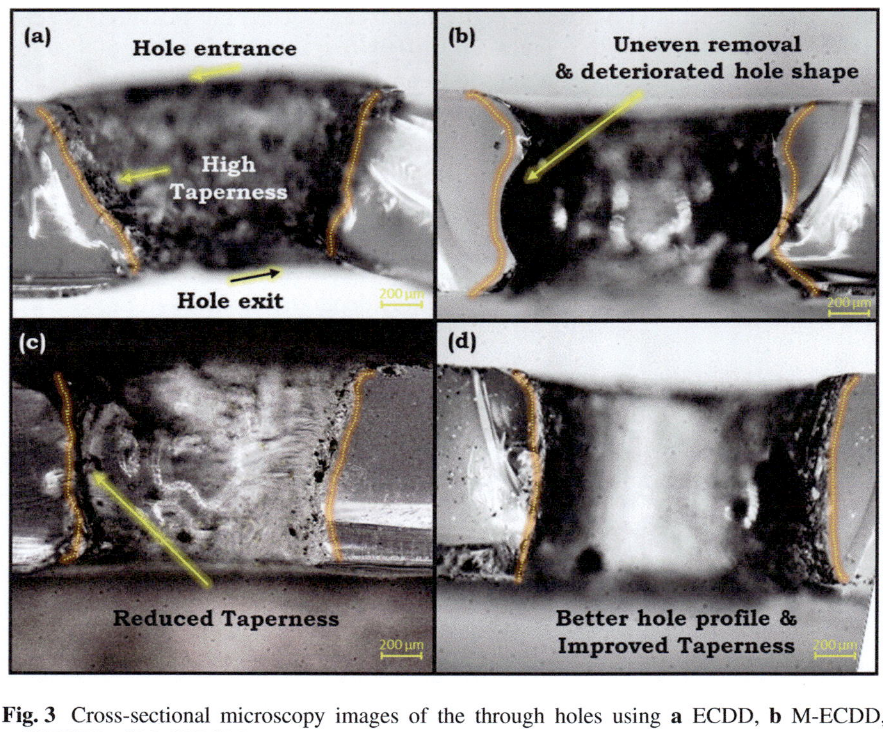

Fig. 3 Cross-sectional microscopy images of the through holes using **a** ECDD, **b** M-ECDD, **c** CLECDD, **d** M-CLECDD processes

in tool contacts in adaptive tool feed further improves the hole characteristics during drilling process.

An improvement of 43.14% and 23.38% is observed in HT in comparison to ECDD and M-ECDD processes, respectively (Fig. 3a–c). From figure, it is clearly visible that the combination of the both the adaptive tool feed with magnetic field in ECDD process (M-CLECDD) produces superior results in comparison to ECDD, CLECDD, and M-ECDD processes (Fig. 3d).

Figure 4 shows the surface roughness graphs of the holes drilled using ECDD and M-CLECDD processes. It is observed that holes with superior surface quality are produced when drilled using magnetic assistance (Fig. 4a). A thinner, more homogeneous gas film forms more quickly in magnetic assistance because of increased bubble creation around the tool due to improved electrolyte circulation. The drilled hole has a smoother finish as a result of this channelized spark occurrence. Better electrolyte etching below the tool also helps to create a smoother surface (Fig. 4a). Adaptive tool feed reduces tool contact, thereby minimizing unnecessary thermal damage. On the other hand, there are greater thermal damage and deteriorated hole shape is produced when magnetic assistance is absent. Because of the uneven spark patterns, low frequency, and non-uniform gas film distribution, the conventional ECDD technique leaves the micro-hole edge with a poor surface quality (Fig. 4c).

The enhanced circulation of fresh electrolyte in the drilling area aids the etching action leading to the improved surface quality. An improvement of 32.71% and 31.91% is noticed at the hole entrance edge and hole exit surface, respectively (Fig. 4a, c).

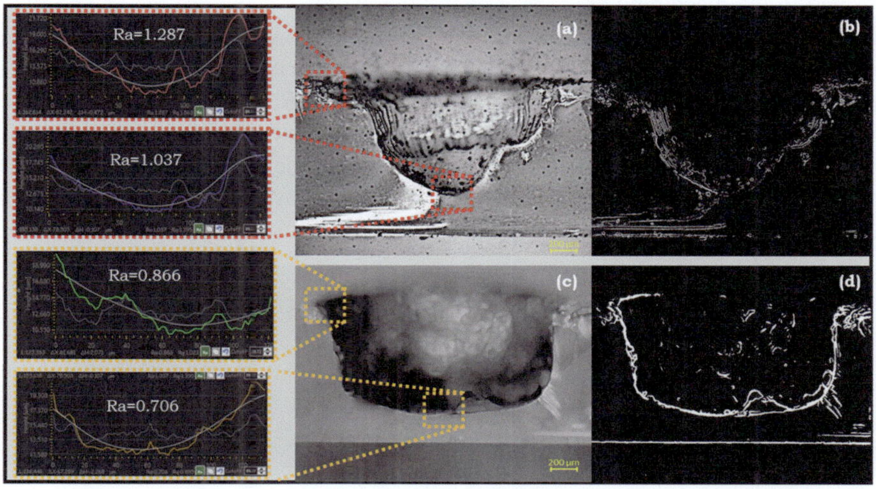

Fig. 4 Cross-sectional microscopy images of the holes along with the surface roughness using **a** ECDD process, **c** M-CLECDD process and glowing line diagram, **b** ECDD process, **d** M-CLECDD process

An increase of 23.89% is observed in MRR with the application of the magnetic field to ECDD process due to increase in electrolyte circulation and uniform spark distribution. The reason for the improvement is seen similar as in case of HT and MD. A similar trend is observed when MRR is compared using M-CLECDD, ECDD, CLECDD, and M-ECDD processes.

3.2 Analysis on the Effect of Different Input Parameters

The effect of tool feed rate, voltage, and concentration is studied on hole characteristics in terms of MD, HT, SR, and MRR. Increase in tool feed rate leads to the decrease in the interaction between the tool and work material which further reduces the material removal leads to the decrease in MRR. High tool feed rate with enhanced electrolyte circulation due to magnetic effect leads to decrease in the thermal energy interaction and increased tool contacts. This causes the formation of non-uniform and unstable thick gas film at the tool vicinity and generates poor discharge frequency. As a result, an increase in HT, SR, and hole deterioration is observed. Figure 5 shows the cross-sectional microscopy images of the holes at two different machining conditions. It is visible from the figure that increase in tool feed rate decreases the MD and increases the HT. Very low tool feed increases the interaction of thermal energy at the hole entrance leading to increase in HT due to increase in hole entrance diameter.

Increasing the applied voltage significantly increases the Lorentz forces which further increases the MHD effect. It induces the rotational effect to the bubbles formed in the machining area. As a result, the bubble's departure radius and residence duration on the tool diminish, concentrating the spark in a smaller region while increasing discharge frequency. Hence, an increase in voltage during magnetic-assisted ECDD process increases the MD, SR, and MRR. Higher voltages generate more hydrogen bubbles at the tool–electrolyte interface, accelerating gas film formation and increasing spark intensity on the hole entry side of the drilling process. It causes the hole entry diameter to expand resulting in the overcut which further causes

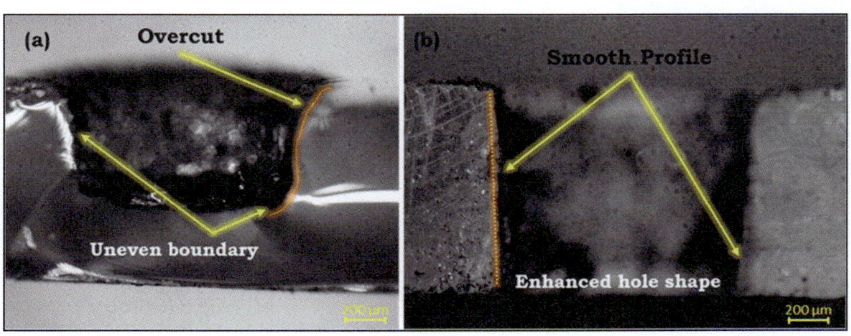

Fig. 5 Cross-sectional microscopy images of the holes drilled using M-CLECDD process at different machining conditions **a** 1 mm/min, 55 V and 30 wt%, **b** 2 mm/min, 45 V and 20 wt%

an increase in HT. Similarly, it is observed that trend of increasing concentration is similar to voltage as it results in the higher MD, increased material removal at the hole entry side, and leads to increase in HT. Overall, MRR improves with the increase in electrolyte concentration. It is accounted due to faster electrochemical processes caused by the quick dissociation of H^+ and OH^- ions, increased spark intensity, and significant heat production in the drilling zone. With the magnetic field assistance, the circulation increases further that enhances the etching action and amplified these effects.

Hence, the selection of the input variables significantly influences the critical hole characteristics, i.e., HT because the desire of improving the MRR and MD with higher voltage and concentration may lead to increase in HT. Overall, it is observed that the low level of tool feed rate with higher value of voltage and concentration produces the worst hole shape (Fig. 5a), leaving behind the selection of levels with utmost care. The medium level of input variables utilized in this study produces the best results keeping in mind the responses, i.e., maximizing MD and MRR and minimizing HT and SR (Fig. 5b)

4 Conclusions

The present study investigates the effect of magnetic field application to the ECDD process using adaptive tool feed system. MD, HT, SR, and MRR are evaluated as a hole characteristic during drilling soda lime glass. Moreover, the effect of input parameters on hole characteristics is discussed. The conclusions are given below:

- Lorenz force is produced in CLECDD with the help of the magnetic field, which further induces an MHD convection. As a result, the electrolyte circulation is improved by the generated MHD convection. The drilled holes' geometrical characteristics, specifically MD and HT, are improved as a result, as MRR.
- The M-CLECDD method exhibits a notable improvement in MD, HT, SR, and MRR of 14.86%, 58.14%, 32.71%, and 23.89% compared to conventional ECDD process.
- Tool contacts and reduced thermal energy interaction provided by high tool feed rate led to unstable gas film development, low discharge frequency, and hole degradation. On the contrary, low tool feed rate increases HT due to more interaction and more material removal at the hole entry side comparative to material removal at hole exit side.
- Increasing voltage improves MRR and MD in the magnetic-assisted ECDD process, but it deteriorates hole taper and roughness because of increased spark intensity and overcut at the hole entrance. Through improved etching and heat generation, the magnetic field amplifies the effects of higher electrolyte concentration, which also enhances MRR, SR, and MD.

References

1. Kumar A, Goud M, Kant S, Rajput V (2024) Synergistic modeling for precision in electrochemical discharge mechanism: a triad of 3D finite element, ANN, and regression approaches. In: Multiscale and multidisciplinary modeling, experiments and design, pp 1–6
2. Ladeesh VG, Manu R (2019) Grinding-aided electrochemical discharge drilling in the light of electrochemistry. Proc Inst Mech Eng C J Mech Eng Sci 233(6):1896–1909
3. Gehlot D, Jha PK, Jain PK (2024) Modelling studies of rotary magnetic field in ECDM for microchannel fabrication of silica glass. Silicon 1–4
4. Gehlot D, Jha PK, Jain PK (2024) Microchannel fabrication on silica glass and experimental investigation of MHD convection in ECDM process. SILICON 16(6):2521–2531
5. Kurafuji H, Suda K (1968) Electrical discharge drilling of glass. Ann CIRP 4(16):415–419
6. Basak I, Ghosh A (1996) Mechanism of spark generation during electrochemical discharge machining: a theoretical model and experimental verification. J Mater Process Technol 62(1–3):46–53
7. Rajput V, Goud M, Suri NM (2021) Multi-response optimization of ECDM parameters for silica (Quartz) using grey relational analysis. SILICON 13:1619–1640
8. Rajput V, Goud M, Suri NM (2021) Review on recent advances, research trends, and gas film in electrochemical discharge-based micromachining. J Micro-and Nano-Manuf 9(1):010801
9. Bhargav KV, Balaji PS, Sahu RK, Katiyar JK (2023) Exemplary approach using tool rotation-assisted μ-ECDM for CFRP composites machining. Mater Manuf Process 38(3):271–283
10. Vemula VV, Chak SK (2020) Effect of loose abrasive mixed electrolyte on the aspect ratio of blind holes drilled in soda-lime glass by ECDM process. Manuf Rev 7:37
11. Hajian M, Razfar MR, Movahed S (2016) An experimental study on the effect of magnetic field orientations and electrolyte concentrations on ECDM milling performance of glass. Precis Eng 45:322–331
12. Gehlot D, Jha PK, Jain PK (2024) Experimental investigation and modelling studies on MHD convection in magnetic-assisted-ECDM. Mater Manuf Process 39(4):494–505
13. Gehlot D, Jha PK, Jain PK (2023) Micro-slit fabrication using magnetic-assisted ECDM and parametric optimization by metaheuristic algorithms. J Braz Soc Mech Sci Eng 45(9):455
14. Xu Y, Jiang B (2021) Machining performance enhancement of deep micro drilling using electrochemical discharge machining under magnetohydrodynamic effect. Int J Adv Manuf Technol 113:883–892
15. Xu Y, Chen J, Jiang B, Liu Y, Ni J (2018) Experimental investigation of magnetohydrodynamic effect in electrochemical discharge machining. Int J Mech Sci 142:86–96
16. Singh M, Dvivedi A, Kumar P (2024) Dimensional accuracy enhancement of machined-hole through UAECDM-process under the magnetic-field-assistance. Mater Manuf Process 39(8):1084–1100

Mechanical Properties of Bound Metal Deposition (BMD) 17-4 PH V2 Materials

James Wheeler, John Belding, and Brett D. Ellis

Abstract Compared to powder-bed fusion (PBF), Bound Metal Deposition (BMD) has substantially less capital and operating costs, yet is safer and has similar performance. Despite these advantages, BMD is less widely utilized than PBF, partially due to a lack of characterization of BMD's process–structure–property relationships. This work seeks to address this problem by characterizing microstructure and mechanical properties of BMD 17-4 PH v2 materials. ASTM E18 hardness and ASTM E8 tensile tests were conducted on 8 XYZ-, 8 YXZ-, and 8 ZYX-oriented (for a total of 24) ASTM E8 subsize tensile specimens. Results indicate the average modulus, yield strength, ultimate strength, percentage elongation, and hardness were 166 GPa, 696 MPa, 887 MPa, 6.33%, and 29.7 HRC, respectively. Average ultimate strengths for the XYZ and YXZ specimens were 941 and 901 MPa, respectively, which were 15% and 10% greater than, respectively, the 820 MPa average ultimate strength of the ZYX specimens. Fracture surfaces were scanned via a z-stacking microscope, revealing fracture morphologies of YXZ and ZYX specimens. Lastly, scanning electron microscopy (SEM) images show the evolution of binder-powder constituents within the microstructure in four states: prior to printing, green state, brown state, and as-sintered. This work is important in that characterization of materials facilitates designers and engineers to realize the aforementioned advantages of BMD.

Keywords Additive manufacturing · Characterization · Bound metal deposition

J. Wheeler
Northeastern University, Boston, MA, USA

J. Belding · B. D. Ellis (✉)
University of Maine, Orono, ME, USA
e-mail: brett.ellis@maine.edu

S. Y. Liang (ed.), *Manufacturing Technologies*, Lecture Notes in Mechanical Engineering, https://doi.org/10.1007/978-3-031-95455-9_2

1 Introduction

Additive manufacturing (AM) is a type of manufacturing which joins material together, often layer by layer, to form a part [1, 2]. As a subsection of AM, additive manufacturing of metal (AMM) is focused on fabrication of metallic material systems and has an estimated 2024 market size of 4.0 billion dollars [3]. The two most commercially common AMM processes, powder-bed fusion (PBF), and directed energy deposition (DED) are large-thermal gradient processes involving melting and re-solidification of monodisperse metallic powders. Although capable of accurate geometries (±0.02 mm), PBF is very expensive and relatively slow with deposition rates of at most 105 cm^3/h [1]. Compared to PBF, DED costs less and has greater deposition rates, but is less accurate (±0.25 mm) [1]. The material extrusion (ME) AMM process, pioneered by Desktop Metal and Markforged in the forms of Bound Metal Deposition (BMD) and Fused Filament Fabrication (FFF), respectively, has reduced costs, and is able to create more homogeneous characteristics and precise parts.

BMD consists of three manufacturing steps: print, debind, and sinter. In Step 1, a two-phase material—metallic powder and a polymer–paraffin–wax binder—is printed by extruding beads into layers to create the desired geometry. The extruded material may be either build material (i.e., desired metallic powder in a binder matrix) or interface (i.e., ceramic powder within a binder matrix). Extruded parts are in the green state and are relatively soft, capable of being deformed via a fingernail. In Step 2, the polymer–wax binder is removed, or debound, via a solvent or a thermal process, thus forming the brown state. BMD v1 materials use solvent debinding; BMD v2 materials use thermal debinding. Lacking binder, brown-state parts are relatively brittle and easy to damage. Whereas solvent debinding requires chemicals, a separate piece of equipment, and relatively long processing times, thermal debinding evaporates the binder in the sintering furnace as an initial segment of thermal cycle within Step 3. In Step 3, the brown-state part is heated to a temperature slightly less than the powdered metal's melting temperature T_m within an argon–hydrogen atmosphere [4]. At temperatures slightly less than T_m, the powdered metal undergoes solid-state densification via atomic diffusion, primarily driven by surface diffusion, grain boundary diffusion, volume diffusion, and evaporation and condensation [5]. The ceramic powder within the interface material neither sinters with the powdered metal nor sinters with itself, thus forming a non-sintered region allowing adjacent metal regions to be easily separated.

The BMD process results in four significant consequences. First, the extrusion process restricts the maximum metallic powder volume fraction, which results in 17–22% linear shrinkage [6]. Second, the sintering process is a low thermal gradient process that neither melts nor re-solidifies powder, thus precluding columnar grain growth common to high-gradient processes such as PBF [7]. Third, the sintering process can utilize wider particle size distribution (PSD) powders, which require less rework during powder manufacture and are therefore cheaper than narrower PSD powders utilized by PBF [4, 7]. Fourth, the extrusion process utilizes all of the feedstock material, thus avoiding issues related to the reusability of non-fused powder [8].

BMD green-state meso-structures consist of at least three components—a raft, an interface, and material for the desired part. If the desired part contains overhangs greater than 45 degrees, support sections may be included to support the brown-state part during sintering. Starting at the bottom, the raft is printed from build material and contains approximately 20–25 layers. Although adjustable between prints and uniform within a print, layer heights of 0.175 mm are typical. On top of the raft, approximately two layers of interface material are printed, thus allowing for easy separation of the raft from the desired part. Depending upon part geometry, the next layers are either support, interface, or the desired part. BMD parts employ an infilled-shell topology in which the outer skin contains five or fewer rasters; regions inside of the outer skin are infilled. The infilled-shell topology locates material where bending stresses are the greatest while reducing wall thicknesses, which reduces the processing time for the diffusion-limited debinding and sintering steps.

17-4 PH is a widely utilized martensitic stainless steel with a typical composition of 0.07% C max, 1.00% Mn, 0.040% P, 0.030% S, 1% Si, 15–17.50% Cr, 3.00–5.00% Ni, and 3.00–5.00% Cu [9]. As a precipitate hardening (PH) steel, 17-4 PH can be hardened via solution treatment for approximately 1 hour at 1,038 °C [1,900 °F], air quenching to less than 32 °C [90 °F], and artificial aged at the "H" hardening temperature listed in Fahrenheit [9]. For example, an "H900" 17-4 PH is artificially aged for 1 hour at 900 °F [482 °C], and H1025 is artificially aged at 1025 °F [552 °C] for 4 hours [9]. The solution treatment and hardening temperatures are less than 17-4 PH's 1404–1440 °C [2560–2625 °F] melting temperature T_m [9].

Table 1 shows manufacturer-stated tensile strength S_u, yield strength S_y, modulus E, % elongation $\% \, el$, hardness H, and relative density ρ_r for BMD v1, BMD v2, FFF, and wrought 17-4 PH materials.

Beyond manufacturer-stated properties, the literature indicates that mechanical properties depend upon process parameters, including specimen orientation [14, 15], layer height and nozzle diameter [16], and sintering batch [4]. For example, Bellezze et al. [14] show that BMD 17-4 PH v1 materials printed into XYZ- and XZY-oriented specimens (*cf.* [17]) were 829 MPa, which was ~8% greater than the 765 MPa for an orientation halfway between XYZ and XZY. Similarly, Bjørheim and La Torraca

Table 1 Mechanical properties of AMM and wrought 17-4 PH stainless steels

Property	As-sintered BMD v1 [10]	As-sintered BMD v2 [11]	As-sintered FFF [12]	Wrought H1025 [13]
S_u (MPa)	1042	925	1180	1070–1270
S_y (MPa)	660	695	710	1000
E (GPa)	195	n/a	152	200
$\% \, el$ (%)	8.5	5.3	7	10
H (HRC)	37	26	36	39
ρ_r (%)	98	n/a	96	100

Lopez [15] show that FFF 17-4 PH materials printed in the ZXY orientation had tensile strengths of 717–738 MPa, which was ~32% less than the 1065–1072 MPa observed for XYZ-oriented specimens. Aside from print orientation, Basak et al. [16] show that layer heights between 0.12 and 0.16 mm and nozzle diameters between 0.25 and 0.40 mm of BMD 316L v1 materials can affect mechanical properties, e.g., 210 MPa $\leq S_y \leq$ 402 MPa and 79 $\pm$ 0.2 $\leq \rho_r \leq$ 97 $\pm$ 0.2.

This work seeks to build upon prior research by characterizing structure and mechanical properties of BMD 17-4 PH v2 materials which is important in that it forms the basis for design decisions and proper utilization of this technology.

2 Methods

The geometry for ASTM E8 [17] 100 $\times$ 10 $\times$ 6 mm^3 subsize tensile specimens was drawn in SolidWorks 2022. The SolidWorks drawing was imported into Desktop Metal (DM) Fabricate v3.2.5 for orientation and slicing into XYZ, YXZ, and ZYX (cf. [18]) print files. The XYZ and YXZ print files each included a single specimen; the ZYX print file included four specimens connected via a top plate and support material between the raft and the top plate. Importantly, the top plate and the support material required by the top plate in the ZYX print file stabilized the specimens during sintering. The XYZ and YXZ print files were each printed eight times on a DM Studio System+ printer, yielding eight XYZ and eight YXZ green-state specimens. The ZYX print file was printed twice, yielding eight green-state specimens. All prints utilized a 0.400 mm diameter print nozzle and a 0.175 mm layer height.

As shown in Fig. 1, YXZ-oriented print file includes 21 raft layers (blue), 2 ceramic interface layers (purple), and 41 print layers (gray and yellow). The inset shows the infill rasters alternating at $\pm 45°$ degrees to the specimen's longitudinal axis. XYZ specimens were similar to the YXZ specimens except that the longitudinal axis was rotated by 90 degrees in the build plane.

ZYX-oriented specimens were printed in two sets of four (cf. Fig. 2), where four specimens were attached via a top plate (not shown), which was removed prior to testing. The print included 21 raft layers (blue), 2 ceramic interface layers (not shown), 691 print layers (gray and orange), and 62 layers for the top plate (not shown). Specimens contained 1 perimeter raster (gray), surrounding 7 rectangular, concentric infill rasters (orange). The positive z surfaces show (1) the outer perimeter raster (gray), (2) inner rasters (orange) following the contours of the outer perimeter raster, and (3) the inner rasters' start-stop points (gray) that form a line from the positive-x, negative-y corner of each layer to the center of each layer.

Green-state parts were thermally debound and sintered in multiple batches in a DM Studio System+ Furnace for approximately 115 h (XYZ and YXZ specimens) and 175 h (ZYX specimens). After sintering the YXZ and ZYX specimens, the furnace was unavailable for approximately 1 month to replace a failed vacuum pump. After repairing the furnace, the XYZ specimens were sintered. Post-processing included

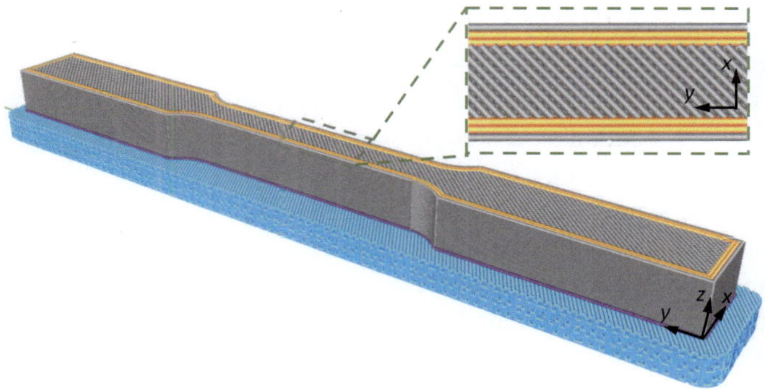

Fig. 1 Raster strategy for YXZ subsize tensile specimen. The perimeter consists of one outer raster (gray) and two inner rasters (orange)

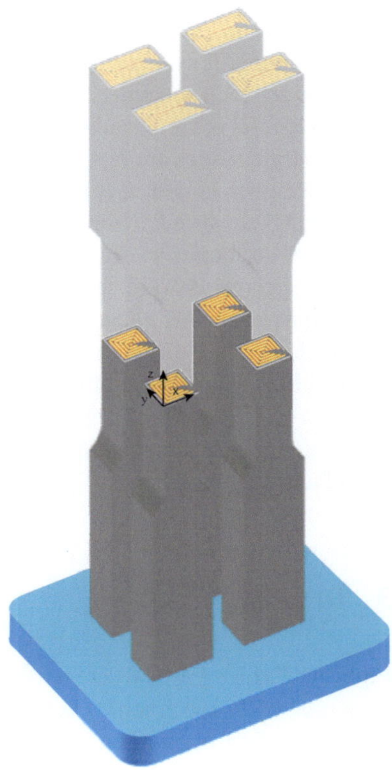

Fig. 2 Raster strategy for ZYX tensile specimens showing concentric layers filling the gauge length. Upper portion of the specimen is shown translucent for clarity

removing specimens from rafts, and removing the ZYX specimens from the top plate via wire electrical discharge machining (EDM).

Prior to tensile testing, the height and width of each specimen's gauge section was measured at three locations—center and near the two ends of the gauge section—with calipers. Each specimen's measured dimensions were averaged to determine the specimen's initial average height, width, and cross-sectional area A_0. Additionally, Rockwell hardness tests were performed per ASTM E18 [19] on the grip section of each specimen. Each specimen was tested 13 times, with the average hardness for each specimen calculated from the last 10 measurements.

Tensile testing was then performed in accordance with ASTM E8 [17] on an MTS C43.504 50-kN load frame using an LPS.504 50-kN load cell. The load frame utilized displacement control and imposed a monotonic displacement rate of 0.008 mm/s. Specimen elongation was measured with an MTS 634.11F-24 extensometer having a 25.0 mm gauge length. All tensile tests were conducted at room temperature.

The modulus of elasticity was calculated using stress–strain values between 0.01% and 0.1% strain. Modulus was used to calculate a 0.2% offset yield strength S_y for each specimen. Tensile strength S_u was calculated as the maximum engineering stress observed for each specimen. All stress values are reported as engineering stress $\sigma = F/A_0$, where F is the instantaneous force divided by the initial cross-sectional area A_0. All strain values are reported as engineering strain $\varepsilon = \delta/L_0$, where $\delta = L - L_0$ is the change in length, L is the instantaneous length, and L_0 is the initial length.

Imaging consisted of optical fractography and scanning electron microscopy (SEM) of structures at various processing steps. Fracture surfaces of ASTM E8 fracture specimens were scanned with a Keyence VHX-7000 optical microscope. Operating in a z-stacking mode at a 200× magnification, the VHX-7000's 4 K camera recorded images at different depths, which were merged to create a composite image that resolved features at multiple heights and measured the heights of those features. Lastly, a Zeiss NVision 40 SEM operating at high tension accelerating voltage of 3.00 kV imaged uncoated specimens at various processing steps throughout the manufacturing workflow.

3 Results

Eight ASTM E8 [17] subsize tensile specimens per orientation were manufactured for a total of 24 specimens. Each specimen was uniquely labeled with a three-letter code indicating orientation per ASTM 52,909 [18] and a one-digit number indicating the specimen number for the orientation. For example, the specimen "XYZ 3" was the third specimen printed such that the X-direction was the major axis of the specimen, the Y-direction was intermediate axis, and the Z-direction was the minor axis. Specimen dimensions were measured at the center and both ends of the gauge section using calipers. Results shown in Table 2 indicate the specimens were slightly larger than ASTM E8's width specification of 6.0 ± 0.1 mm [17]. Here, height of the gauge section is the dimension in the minor direction (e.g., dimension in the *Z*-direction for an XYZ specimen), and width of the gauge section is the dimension in the intermediate direction (e.g., dimension in the *Y*-direction for an XYZ specimen). Table 2 also shows mean and sample standard deviation for hardness by orientation.

The engineering stress–strain curves and mean and sample standard deviation values for the 24 specimens are shown in Fig. 3 and Table 3, respectively. XYZ specimens had the greatest mean yield strength with 720 MPa and greatest mean tensile strength with 941 MPa. Whereas ZYX specimens had the greatest mean modulus at 172 GPa, YXZ specimens had the greatest mean percentage elongation at 8.0%. Including all three orientations, the overall means for modulus, yield strength, ultimate strength, and percentage elongation were 166 GPa, 696 MPa, 887 MPa, and 6.3%, respectively.

After tensile testing, fractography was conducted via a VHX-7000 z-stacking microscope. Fractography indicated a column of voids on many of the XYZ and YXZ specimens' fracture surfaces, as shown in Fig. 4a , b, for the representative specimen YXZ 6. Figure 4a, b also shows that the center of the fractured surface is lighter colored and at a greater elevation, respectively, than the outer region, potentially indicating different failure mechanisms.

As indicated in Fig. 4c for the representative specimen ZYX 3, the fracture surfaces for the ZYX specimens were qualitatively different than those of the XYZ and YXZ specimens. In Fig. 4c, the fracture predominantly occurred at interfaces between

Table 2 Gauge section dimensions and hardness by orientation

Orientation	Statistic	Height (mm)	Width (mm)	Area (mm^2)	Hardness (HRC)
XYZ	Mean	6.21	6.34	39.37	33.3
	StDev	0.02	0.02	0.20	2.3
YXZ	Mean	6.24	6.39	39.90	28.4
	StDev	0.02	0.02	0.19	1.8
ZXY	Mean	6.37	6.29	40.03	27.5
	StDev	0.02	0.02	0.17	1.9

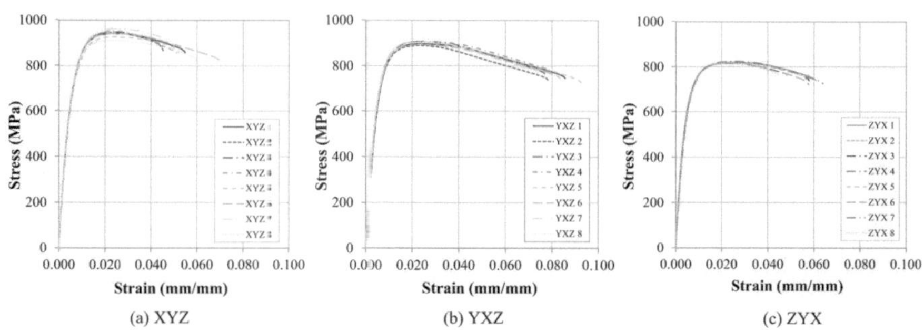

Fig. 3 Stress–strain curves for BMD 17-4 PH v2 in XYZ, YXZ, and ZYX orientations

Table 3 Tensile properties of BMD 17-4 PH v2 tensile specimens by orientation

Orientation	Value	E (GPa)	S_y (MPa)	S_u (MPa)	% el (%)
XYZ	Mean	164	720	941	5.4
	StDev	11.5	12.6	16.0	0.8
YXZ	Mean	161	728	901	8.0
	StDev	19.1	31.6	7.6	0.7
ZYX	Mean	172	640	820	5.6
	StDev	17.4	19.2	2.3	0.6

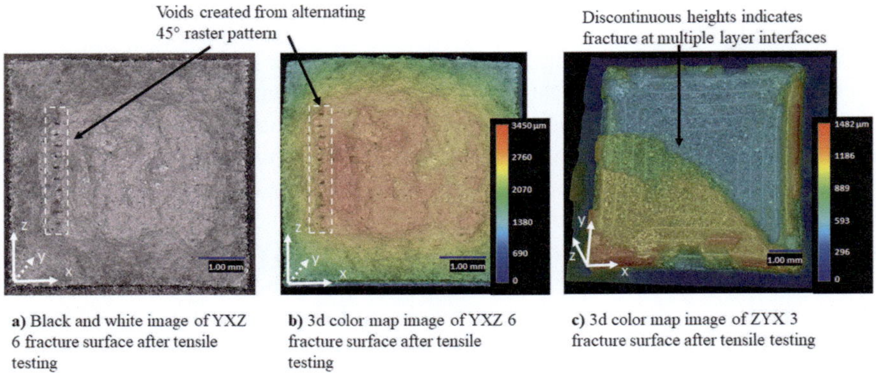

Fig. 4 Keyence VHX-7000 images showing **a** YXZ 6 fracture surface, **b** color map image of YXZ 6 fracture surface, and **c** color map image of ZYX 3 fracture surface

layers connected by abrupt changes between interfaces. For example, Fig. 4c shows a large proportion of the fracture occurred in the top-right blue-colored region, which is adjacent to the green-colored region representing a different print layer. The blue- and green-colored regions have an abrupt change in height (cf. black arrow) between them. In Fig. 4c, the white-in-appearance, square-shaped lines indicate inter-raster voids. Noting that the abrupt changes in interfaces are predominantly at 45° (bottom right) or 90° (mid-height left) to the inter-raster voids and recalling ZYX specimens have stop-start points from the positive-x-negative-y corner to the center of the cross section (cf. gray line within the yellow inner perimeter rasters in Fig. 2), it is likely the start-stop points for the inner perimeter rasters served as fracture initiation sites.

In addition to 3D microscopy, Scanning Electron Microscopy (SEM) imaging was completed to improve understanding of the structural characteristics of the material. SEM images were taken of a rod-shaped piece of feedstock (i.e., prior to printing, cf. Fig. 5a), on the positive z-face of an as-printed cube (i.e., prior to debinding and sintering, cf. Fig. 5b, c), on the positive z-face of a cube after thermal debinding (cf. Fig. 5d, e), and on the side of a sintered cube part (cf. Fig. 5f). The pre-debinding images (cf. Fig. 5a–c) show the polymer–wax binder as a dark matrix between lighter colored metal particles. The dark-colored polymer–wax binder is predominantly absent in thermally debound structures shown in Fig. 5d, e. In Fig. 5f, individual particles have fused together to form the as-sintered microstructure.

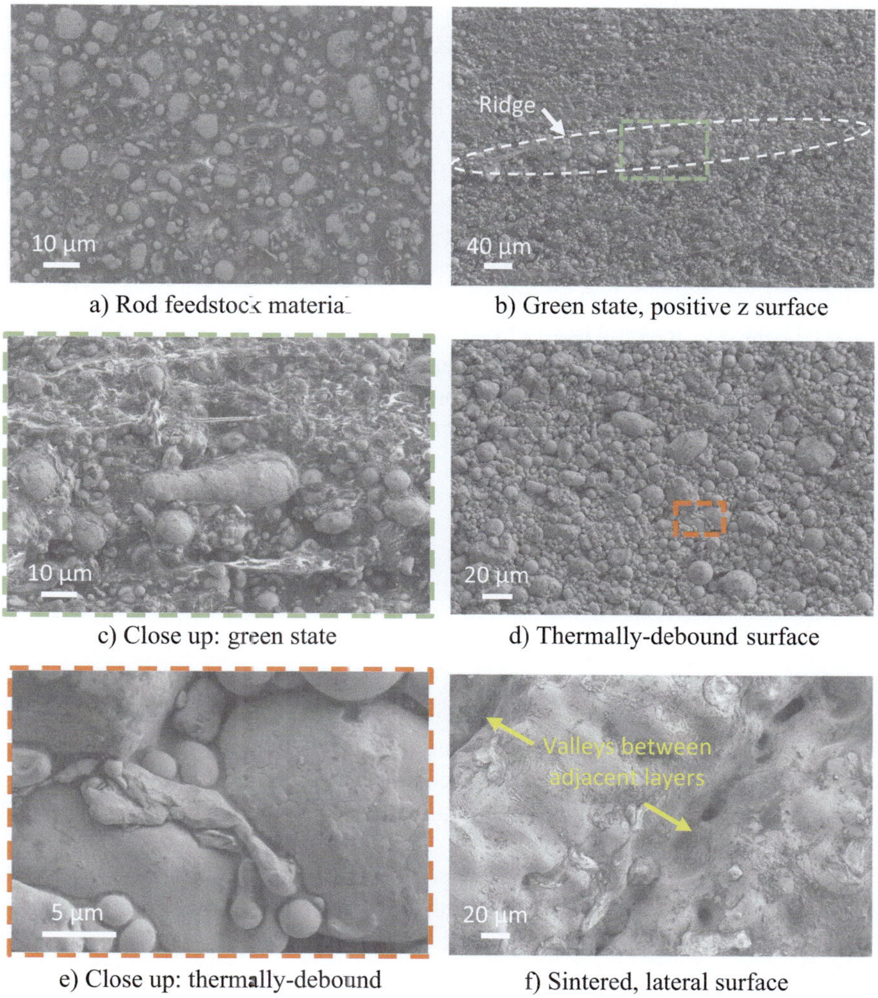

Fig. 5 Scanning electron microscope (SEM) images of BMD 17-4 PH v2 microstructures at progressive manufacturing steps: **a** rod feedstock material, **b** top surface in the green state, **c** close-up of the top surface in the green state, **d** thermally debound surface, **e** close-up of the thermally debound surface, and **f** side of a sintered cube, showing three layers and the two valleys formed between the three layers

4 Discussion

An indirect comparison of BMD 17-4 PH v1 and v2 mechanical properties is possible by comparing results from Watson et al. [4] to the XYZ and YXZ results presented in Sect. 3. For a total of 19 XYZ and YXZ BMD 17-4 PH v1 specimens, Watson et al. [4] report a 176 GPa average modulus, 604 MPa yield strength, 776 MPa tensile strength, and a 6.7% elongation. Of the 24 BMD 17-4 PH v2 specimens characterized in Sect. 3, 16 specimens (i.e., 8 XYZ and 8 YXZ specimens) had in-plane orientations. The 16 in-plane-oriented specimens had 162 GPa average modulus, 724 MPa yield strength, 921 MPa tensile strength, 6.7% elongation, and a 30.9 HRC hardness. Hence, the v2 material had an 8% reduced modulus, 20% greater yield strength, 19% greater ultimate strength, and the same percentage elongation as compared to the v1 materials. Based upon the 24.7–27.2 HRC range of hardness values in Watson et al. [4], the 17-4 PH v2 material had a 13.6–25.1% greater hardness. Despite Watson et al. [4] and this work utilizing the same printer and furnace, the comparison between v1 and v2 materials is indirect due to the approximate 4-year time span between sets of experiments and the resulting possible changes in software, equipment performance, and sintering batches.

5 Conclusions

Desktop Metal's Bound Metal Deposition (BMD) additive manufacturing process was used to manufacture 24 ASTM E8 [17] subsize 17-4 PH v2 tensile specimens. Specimens were characterized via measurement of gauge dimensions, hardness testing of grip regions, tensile testing, and microscopy of fractured surfaces. Scanning electron microscopy was also conducted on specimens prior to printing, prior to thermal debinding (green state), after thermal debinding (brown state), and after sintering.

Results indicate average mechanical properties of 166 GPa modulus, 696 MPa yield strength, 887 MPa ultimate strength, 6.33% elongation at failure, and 29.7 HRC hardness. These results indicate that BMD 17-4 PH v2 has similar mechanical properties and structural characteristics to that of BMD 17-4 PH v1, albeit with a significantly more streamlined manufacturing process.

This work expands upon the collection of work which has covered BMD 17-4 PH v1's chemical debinding, bringing to light the differences between these two versions of the process. This sets a backbone for understanding BMD's thermal debinding, and will facilitate engineers and designers to make design decisions moving forward.

Acknowledgements This research was funded by the University of Maine's Advanced Manufacturing Center. The authors thank Dustin Jensen for help with printing and sintering, Emma Perry and the University of Maine's Electron Microscopy Laboratory for support and help with SEM imaging, and Keith Berube for assistance with tensile testing.

References

1. Ngo TD et al (2018) Additive manufacturing (3D printing): a review of materials, methods, applications and challenges. Compos Part B Eng 143:172–196
2. ASTM (2015) Standard terminology for additive manufacturing - General principles – Terminology. ASTM 52900, West Conshohocken, PA
3. Precedence Research, Metal additive manufacturing market size, share, report 2024–2034. https://www.precedenceresearch.com/metal-additive-manufacturing-market. Last accessed 26 August 2024
4. Watson A, Belding J, Ellis BD (2020) Characterization of 17-4 PH processed via bound metal deposition (BMD). In: 149th annual meeting and exhibition supplemental proceedings of the minerals, metals and materials society (TMS 2020). Springer, Switzerland, pp 205–216
5. German R (2005) Powder metallurgy and particulate materials processing. Metal Powder Industry, Princeton, NJ
6. Desktop Metal, Deep dive: Bound Metal Deposition (BMD). https://www.desktopmetal.com/resources/deep-dive-bmd. Last accessed 26 August 2024
7. Das S, Bourell DL, Babu SS (2016) Metallic materials for 3D printing. MRS Bull 41(10):729–741
8. Moghimian P et al (2021) Metal powders in additive manufacturing: a review on reusability and recyclability of common titanium, nickel and aluminum alloys. Add Manuf 43:102017
9. KVA Stainless, 17-4 stainless steel. http://www.kvastainless.com/17-4.html. Last accessed 19 August 2024
10. Desktop Metal, 17-4 PH stainless steel. https://s3.amazonaws.com/desktopmetal/Studio-MDS-17-4-PH-stainless-steel.pdf. Last accessed 31 December 2018
11. Desktop Metal, 17-4 PH stainless steel PureSinter furnace. https://www.desktopmetal.com/uploads/20240514_MDS_Studio2_PureSinter_17-4PH.pdf. Last accessed 01 August 2024
12. Markforged, Material datasheet: 17-4 PH stainless steel v2. https://www-objects.markforged.com/craft/materials/17-4PH-v2-SS-Datasheet.pdf. Last accessed 26 July 2022
13. Avallone EA, Baumeister III T, Sadegh AM (eds) (2007) Marks' standard handbook for mechanical engineers, 11th ed. McGraw-Hill, New York, NY
14. Bellezze T, Forcellese A, Forcellese P, Mancia T, Simoncini M (2023) Effect of printing orientation on mechanical properties of components in stainless steel obtained using the bound metal deposition technology. Procedia CIRP 118:694–698
15. Bjørheim F, La Torraca Lopez IM (2021) Tension testing of additively manufactured specimens of 17-4 PH processed by bound metal deposition. Mater Sci Eng 1201(1):012037
16. Basak AK et al (2024) Bound metal deposition of stainless steel 316L: effect of process variables on microstructural and mechanical behaviors. Materialia 36:102196
17. ASTM (2022) Standard test methods for tension testing of metallic materials. ASTM E8/E8M, West Conshohocken, PA
18. ASTM (2015) Additive manufacturing of metals - Finished part properties - Orientation and location dependence of mechanical properties for metal powder bed fusion. ASTM 52909, West Conshohocken, PA
19. ASTM (2022) Standard test methods for Rockwell hardness of metallic materials. ASTM E18, West Conshohocken, PA

Hybrid Intelligent Algorithm for Modeling and Optimization of Laser Microgrooving of Sapphire

Ali Naderi Bakhtiyari, Kongbo Wang, Zhenghao Wang, Liyong Wang, Yongling Wu, and Hongyu Zheng

Abstract The emergence of intelligent techniques has revolutionized traditional approaches in manufacturing. This study explores the effectiveness of a hybrid intelligent method for modeling and optimizing the laser microgrooving process on sapphire. It focuses on examining the impact of three critical laser machining factors (laser fluence, scanning speed, and number of passes) on the geometric properties of the resulting microgrooves. Initially, the experimental dataset is used to train and test Artificial Neural Networks (ANN). Bayesian Optimization Algorithms (BOA) are implemented to fine-tune the hyperparameters of ANN to enhance the accuracy of the model. The results demonstrate the model's robust capability to predict the depth and width of laser-microgrooved sapphire, providing a deeper insight into the process dynamics. In the subsequent phase, the study leverages the robust modeling prowess of the ANN-BOA and the impressive efficiency of the Improved Grasshopper Optimization Algorithm (IGOA) to produce microgrooves with optimal aspect ratios. In this regard, the well-trained ANN-BOA models are utilized as objective functions in the IGOA. This integration results in optimal processing conditions for fabricating microgrooves with superior aspect ratios. The results show that optimal laser parameters of 3.49 J/cm^2 fluence, 544 mm/min scanning speed, and 5 passes can achieve microgrooves on sapphire with a favorable aspect ratio of about 2, demonstrating the efficacy of the hybrid ANN-BOA-IGOA optimization approach. Overall, the study highlights the potential of hybrid intelligent methods in modeling and optimizing complex manufacturing processes.

Keywords Laser microgrooving · Multi-objective optimization · Artificial neural network

A. N. Bakhtiyari (✉) · K. Wang · Z. Wang · L. Wang · Y. Wu · H. Zheng (✉)
Centre for Advanced Laser Manufacturing (CALM), School of Mechanical Engineering, Shandong University of Technology, Zibo, P. R. China
e-mail: ali.naderi.bakhtiyari@outlook.com

H. Zheng
e-mail: y111zheng@163.com

S. Y. Liang (ed.), *Manufacturing Technologies*, Lecture Notes in Mechanical Engineering, https://doi.org/10.1007/978-3-031-95455-9_3

1 Introduction

Sapphire is an advanced ceramic that is extensively employed in crafting crucial components due to its exceptional characteristics [1]. Sapphire has electrical insulation capability, elevated hardness, a minimal thermal expansion coefficient, resistance to corrosion, and the ability to withstand high temperatures [2]. It is noteworthy that sapphire substrate poses a significant challenge for cutting due to its extreme brittleness [3].

Consequently, achieving precise machining of sapphire components becomes a complex endeavor considering these intricacies. The fabrication of microstructures such as grooves and holes, particularly those with sub-millimeter dimensions, is problematic using conventional mechanical methods. Consequently, numerous researchers have shifted their focus to non-conventional techniques for the precise machining of sapphire [4, 5]. This shift aims to circumvent the issues of tool breakage and wear inherent in mechanical methods, with a significant emphasis now placed on non-contact approaches.

Among non-conventional methods, laser machining has gained considerable interest due to its additional benefits [6, 7]. These include high geometric flexibility and the ease of crafting complex microstructures through the controlled scanning of the laser beam [8]. Pulsed lasers, especially, have proven efficient for machining hard-to-cut materials thanks to their high peak power and low pulse width [9, 10].

However, in laser machining of sapphire, accumulating sufficient energy for effective ablation is challenging due to sapphire's large bandgap [11]. To address this issue, Bakhtiyari et al. [12] employed a silicon–sapphire interface absorption technique for machining sapphire. In this approach, a silicon plate was placed beneath the sapphire surface to utilize the generated plasma as a source of heat and pressure for machining. This technique was found to significantly enhance the absorption of a nanosecond (ns) pulsed laser with a wavelength of 1030 nm (nm) by sapphire.

In the laser machining process, different machining input processing parameters such as laser power, pulse width, laser fluence, scanning speed, and spot size play a crucial role [13–15]. These parameters align closely with the derived energy equation of heat transfer, making them highly compatible and suitable for achieving precise results. Typically, the processing parameters are adjusted and employed in machining to achieve the desired quality of microgrooves [16]. However, this process can be time-consuming, and there is still room for researchers to model the behavior of the process and identify optimal conditions for high-quality machining [17]. In response to this shortcoming, this study employed a blend of intelligent methods to anticipate the geometric properties of nanosecond laser-machined sapphire and identify the most effective processing parameters for producing a microgroove with a high aspect ratio. The proposed Artificial Neural Network (ANN)–Bayesian Optimization Algorithm (BOA) model in our research not only seeks to understand and predict the dynamics of manufacturing processes but also aims to fulfill the overarching goal of the Improved Grasshopper Optimization Algorithm (IGOA). Essentially, we leveraged the capabilities of ANN for modeling complex behaviors to define cost functions in the context of multi-objective grasshopper optimization.

2 Laser Microgrooving of Sapphire via Silicon–Sapphire Interface Absorption

Figure 1 illustrates the intricate laser machining process occurring at the interface between the silicon (Si) target and the sapphire substrate. This process encompasses a series of complex phenomena, beginning with the interaction of the laser beam with the Si target. Initially, the Si target absorbs the laser energy, leading to its heating and ionization. This absorption and subsequent ionization are critical in forming a plasma, a hot, ionized state of matter comprising vapors, liquid droplets, and solid particles ejected from the Si target [12]. As this plasma forms, it begins to rise and expand toward the rear surface of the sapphire substrate. This expansion is isothermal and is significantly influenced by a process known as inverse bremsstrahlung (IB) absorption, which occurs during the laser–plasma interaction. IB absorption refers to the absorption of the laser energy by the plasma, leading to its expansion and heating. The plasma plume's interaction with the sapphire substrate plays a pivotal role in the microgrooving process. It enhances the transient absorption of the laser energy by the substrate, facilitating the precise and controlled removal of sapphire material. This interaction leads to the formation of microgrooves with specific geometric properties, determined by the dynamics of the laser, plasma, and substrate interaction.

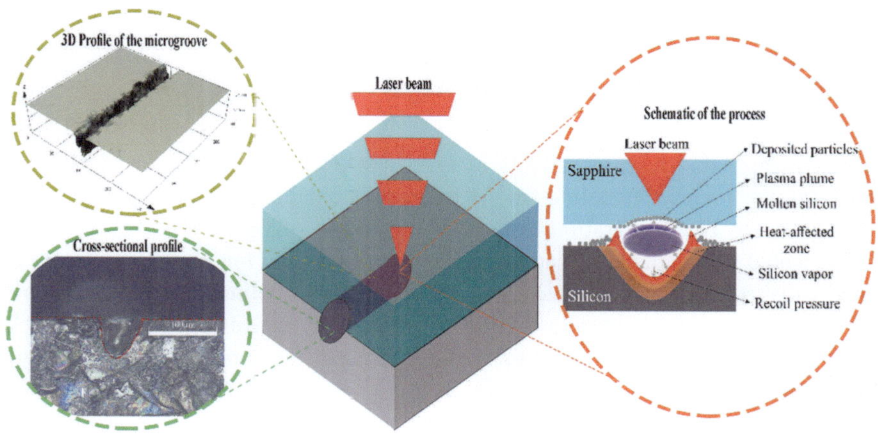

Fig. 1 The laser microgrooving of sapphire via Si–sapphire interface absorption

Table 1 Processing parameters and their respective levels used in the full factorial experiment

Parameters	Unit	Levels
Laser fluence	J/cm^2	2.8, 3.2, 3.6
Scanning speed	mm/s	100, 250, 400, 550
Number of passes	–	1, 2, 3, 5, 8, 13

3 Experiment

The experiment utilized a 500-μm-thick sapphire (C-plane) wafer as the substrate material, paired with a silicon (type P) wafer of identical thickness as the target material. The laser system employed had a pulse duration of 2 ns and a wavelength of 1030 nm, precisely focused at the Si–sapphire interface.

Key process parameters were carefully selected for their significant impact on the laser microgrooving outcomes, particularly affecting the depth and width of the microgrooves. The specific values and levels for these processing parameters are presented in Table 1. By manipulating these parameters, the experiment involved a multilevel factorial design approach, enabling the creation of 72 distinct microgrooves on the sapphire substrate.

The sapphire substrate and Si target underwent a thorough cleaning procedure, including detergent washing, water rinsing, and alcohol immersion in an ultrasonic bath, both before and after the process. Furthermore, the sapphire substrate underwent a 20-min etching treatment in hydrochloric acid to remove any deposited particles on the samples. The morphology of the machined microgrooves was assessed using a 3D profilometer (Rtech-UP24).

4 Modeling and Optimization Algorithms

In this regard, this research adopts a novel approach, integrating an ANN with BOA for the precise modeling of the geometric characteristics of laser-induced microgrooves. This integration leads to the creation of the ANN-BOA model, a sophisticated tool capable of capturing the nuances of the microgrooving process. Once fully trained and validated, the model outputs, delineating the geometric features of the microgrooves, are utilized as the objective functions in the IGOA. The methodology encompassing the ANN-BOA and the subsequent optimization process is methodically outlined in Fig. 2.

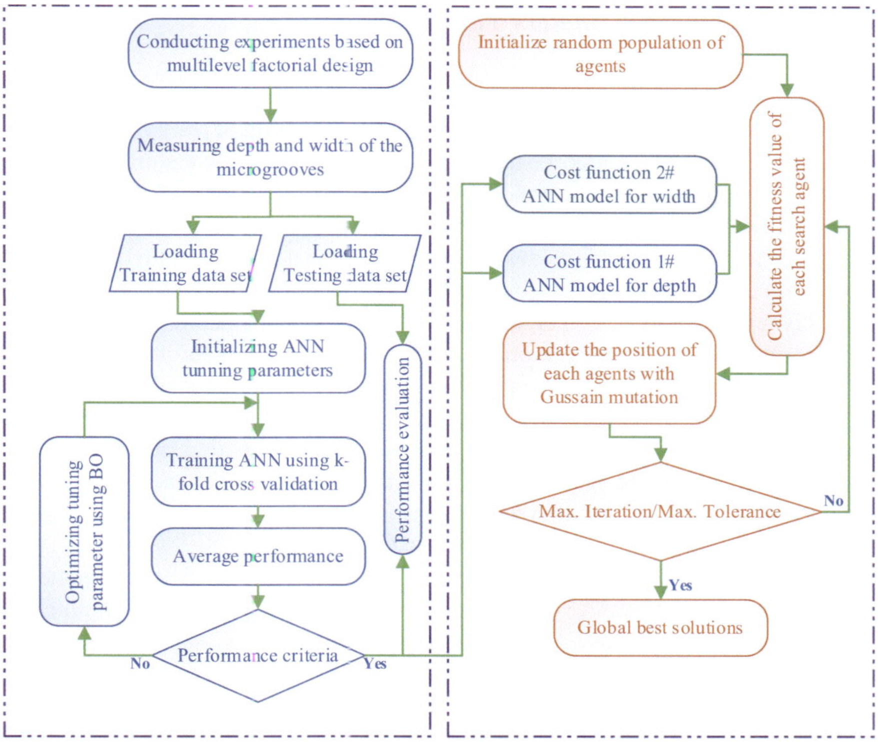

Fig. 2 Flowchart of the proposed hybrid intelligent technique (ANN-BOA-IGOA)

4.1 Modeling Using Artificial Neural Network

ANN has gained widespread recognition for its remarkable ability in function approximation and pattern recognition across a variety of datasets [15]. The increasing reliance on ANN can be attributed to their architecture, which mirrors the neural networks of the human brain. This intricate design allows ANN to process complex and nonlinear data with remarkable efficiency, making them particularly adept at modeling diverse types of real-world data [18].

At its core, an ANN is composed of interconnected processing units known as neurons. These neurons, connected through numerical weights, are organized into several layers, functioning similar to nodes within a directed graph. Each neuron operates in two fundamental stages. Initially, it computes the weighted sum of its inputs, adding a bias term for calibration. This sum is then passed through an activation function, which modulates the neuron's output based on a pre-defined algorithm. The cumulative output of the ANN is generated from the outputs of neurons in its final layer, with each neuron's output having been modified by the activation function.

4.2 ANN Hyperparameters Tuning Using BOA

The fine-tuning of its hyperparameters significantly influences the efficacy of an ANN during the learning process. Identifying the optimal set of learning parameters and, in some cases, the network structure itself is crucial for achieving the highest accuracy in predictions. This is typically viewed as an optimization challenge. Traditional trial-and-error methods, while common, tend to be time-intensive and resource-heavy. Consequently, there is a growing inclination toward more intelligent, efficient approaches for hyperparameter optimization.

In this regard, BOA has emerged as a formidable and adaptable tool, especially effective in optimizing complex processes with incomplete or limited information [19]. Their prowess in fine-tuning machine learning hyperparameters renders them invaluable for boosting the performance of ANN models [20]. BOA operates on a probabilistic surrogate model and utilizes an acquisition function, which is instrumental in determining the optimal design points in a complex system. This function critically balances the exploration of new possibilities with the exploitation of known good ones, a balance vital for efficient optimization.

Given the remarkable proficiency of BOA in identifying optimal hyperparameters for ANN, we utilized BOA to determine the most effective ANN hyperparameters. During the Bayesian Optimization Algorithm (BOA) process, this objective function is iteratively optimized by adjusting the ANN hyperparameters, including the number of hidden layers, the activation functions, and the number of neurons, to minimize the MSE. The ultimate goal is to achieve high prediction accuracy for the geometric characteristics of the laser-microgrooved sapphire, which will be used later in the optimization process with IGOA. The optimal tuning parameters identified through the BOA optimization process are detailed in Table 2. This strategic approach underscores BOA's efficacy in enhancing ANN performance.

Table 2 The tuning hyperparameters of ANN-BOA models

Model hyperparameters	Search range	Optimized tuning parameters	
		Depth	Width
Number of hidden layers	1–3	3	2
Regulation strength (Lambda)	1.59e-7–1.59e3	3.89e-5	1.73e-2
Activation function	ReLU, Tanh, Sigmoid	ReLU, ReLU, ReLU	Tanh, Tanh
Number of neurons	1–300	215, 293, 276	12, 290

Table 3 The applied parameters in IGOA

Parameters	Values
Maximum number of iteration	500
Population size	100
Maximum comfort zone (C_{max})	1
Maximum comfort zone (C_{min})	0.00003
Intensity of attraction	0.5

4.3 *Optimization of Laser-Induced Microgroove Using Improved Grasshopper Algorithm*

In recent years, nature-inspired algorithms have been increasingly adopted in scientific and industrial applications, renowned for their straightforward, gradient-free mechanisms, and their robust ability to circumvent local optima [21]. These algorithms are particularly advantageous in their approach to treating complex problems as black boxes, simplifying the optimization process. They draw inspiration from natural problem-solving strategies observed in various creatures, mimicking their survival-driven evolutionary and adaptive techniques. A prominent example is the IGOA, noted for its efficacy and wide-ranging applicability in various domains [22, 23]. The IGOA is inspired by the natural swarming behavior observed in grasshoppers. This algorithm integrates two critical phases: exploration and exploitation [24]. Initially, the algorithm deploys search agents in an exploratory phase to broadly scan the solution space. Subsequently, in the exploitation phase, these agents demonstrate localized movements, focusing on promising areas. This cycle of exploration and exploitation continues iteratively, with positions being updated until a pre-defined termination criterion is achieved. After this process, IGOA identifies the most optimal solution, represented by the position and fitness of the target closest to the global optimum.

The main goal of this study is to maximize the microgroove depth while simultaneously minimizing its width. To this end, the ANN-BOA model was established and employed as the objective function in the optimization process. The initial parameters for IGOA are detailed in Table 3.

5 Results and Discussion

Evaluating the performance of the trained models is a critical step in establishing their predictive accuracy and general applicability. A key aspect of this evaluation is testing the model against data that were not involved in its training phase, ensuring the model's robustness and generalizability. For this purpose, a subset of nine datasets was randomly selected from the total pool of 72 for a comprehensive evaluation. The performance of the developed ANN models, both on the training and testing datasets, was analyzed and is summarized in Table 4. This table presents the values of two critical performance metrics: Root Mean Square Error (RMSE) and Mean Absolute Error (MAE), for both depth and width predictions of the laser microgrooves. These metrics provide insights into the excellent precision of the model predictions. Moreover, the substantial agreement between the training and testing data results, as evidenced by the RMSE and MAE values, underscores the model's robustness.

Figure 3 shows regression plots comparing the predicted attributes of laser-microgrooved sapphire with the actual experimental results, as determined by the ANN model. The close alignment between the predictions and the experimental data validates the accuracy of the models and underscores their robust predictive capabilities. This strong correlation highlights the effectiveness of the ANN-BOA model in capturing and predicting the complex geometric properties of the microgrooves. As previously demonstrated, the developed models exhibit excellent accuracy in predicting the geometric characteristics of laser-induced microgrooves. These models allow for predictions across a range of processing parameters, offering valuable insights into the intricate relationships within the process. Figure 4 illustrates these insights by showing variations in the depth and width of microgroove samples based on different combinations of laser fluence and scanning speed. A clear trend emerges: as the scanning speed increases, both the depth and width of the microgrooves decrease. This reduction occurs because higher speeds result in fewer laser pulses per unit area, lowering the total energy transferred to the material and leading to shallower grooves. Conversely, slower scanning speeds increase the overlap of laser pulses, causing more intense heating and greater melting and evaporation of the substrate, resulting in deeper and wider grooves. Additionally, a higher repetition rate of laser pulses enhances the interaction with the Si target material, leading to a greater thermal impact. When combined with higher laser fluence, this effect deepens and widens the microgrooves. Pulse energy also plays a critical role in this process. Higher pulse energies create stronger recoil pressure, which significantly boosts the ablation rate through mechanical effects.

It is well established that the efficacy of optimization is inherently linked to the precision of the cost functions used. In this regard, the ANN-BOA models, as established earlier in the study, were utilized to inform the IGOA accurately. As shown in Fig. 5a, the resulting Pareto-optimal front is a visual representation of the equilibrium between depth and width in the microgrooves, achieved through our optimization efforts. The Pareto-optimal front's specific formation is attributed to the continuous nature of the optimization challenge. This Pareto front maps out a continuum of

Table 4 The value of performance functions for prediction of depth and width of the microgrooves using ANN-BOA models

	Depth		Width	
	Training data	Testing data	Training data	Testing data
RMSE	2.17	2.52	0.78	2.17
MAE	165	1.91	0.62	1.46

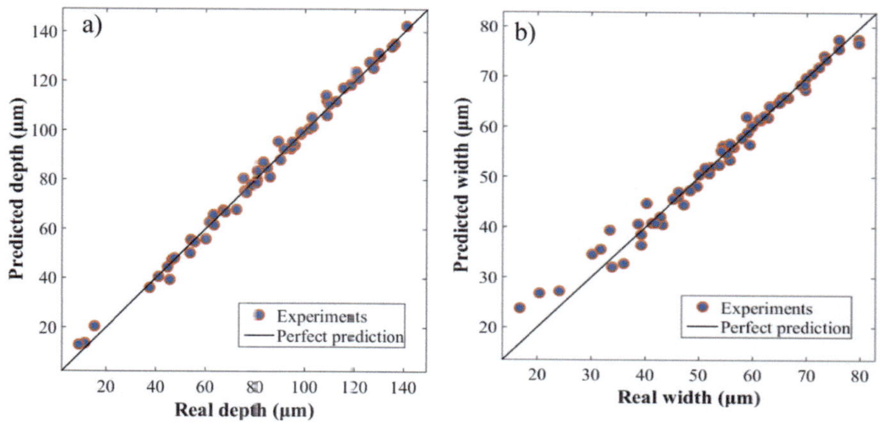

Fig. 3 Regression plot for the proposed models in predicting **a** depth and **b** width of the microgrooved sapphire

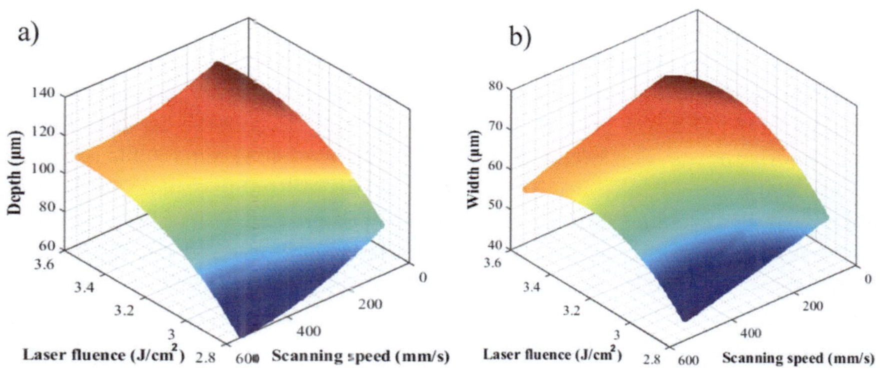

Fig. 4 The predicted **a** depth and **b** width of the machined sapphire using ANN methods for four times laser passes

feasible solutions, each corresponding to a set of process parameters that deliver optimal outcomes. Selecting the most appropriate set from this continuum is influenced by specific production goals, engineering discretion, operational constraints, and resource availability.

Figure 5b illustrates a representative microgroove on a sapphire substrate, achieved under optimized laser machining conditions. The machining was performed at a laser fluence of 3.49 J/cm^2, a scanning speed of 544 mm/min, and involved 5 passes. The graph showcases the microgroove's depth and width, measured at approximately 109.8 μm and 57.9 μm, respectively, resulting in an aspect ratio close to 2. This finding highlights the effectiveness of the proposed technique in achieving superior geometric proportions in the laser microgrooving of sapphire.

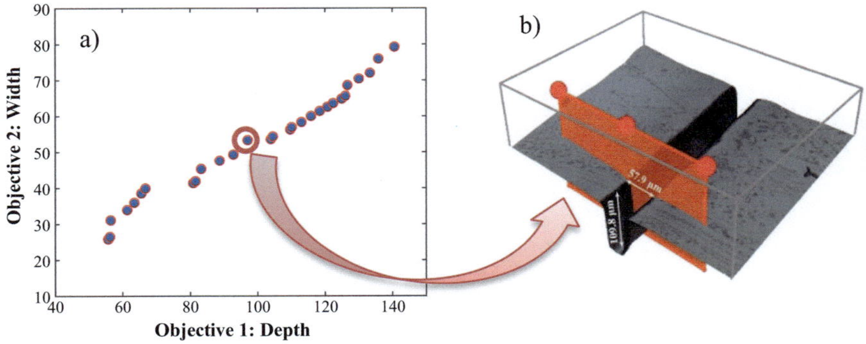

Fig. 5 **a** Pareto front for multi-objective optimization of the laser-induced microgrooves. **b** 3D representation of a laser-microgrooved sapphire substrate at optimal parameters

6 Conclusion

This study aimed to enhance the precision and effectiveness of laser microgrooving on sapphire substrates by combining the strengths of ANN, BOA, and IGOA. Initially, the ANN model was rigorously trained and tested using hyperparameters finely tuned by BOA. This was done to ensure precise predictions of the depth and width of the microgrooves. This predictive model has demonstrated remarkable accuracy, as indicated by low RMSE and MAE values. It was utilized to explore further and understand the dynamics of the laser microgrooving process. It enabled predictions across a range of processing parameters, offering valuable insights into the intricate relationships within the process. Additionally, the effective modeling abilities of the developed ANN-BOA were synergistically combined with the advanced multi-objective optimization capabilities of IGOA. This integration aimed at optimizing the laser microgrooving process, leveraging both techniques' strengths in handling complex datasets and multiple objectives. At this stage, the IGOA was utilized to concurrently maximize the depth and minimize the width of microgrooves, ensuring a high aspect ratio. The Pareto front showcases a range of optimal solutions that effectively balance the depth and width of microgrooves. It presents a variety of processing options, each uniquely suited to meet specific manufacturing requirements. The dual applications of modeling and optimization not only underline the efficacy of hybrid intelligent algorithms but also open up new avenues for their application in the manufacturing sector, promising a significant impact on global market competitiveness.

References

1. Dobrovinskaya ER, Pishchik V, Lytvynov LA (2009) Sapphire: material, manufacturing, application. Springer

2. Berndt M, Danner M (2016) Laser Tech J 13:42
3. Xie XZ, Zhou CX, Wei X, Hu W, Ren QL (2019) Opto-Electron Adv 2:1
4. Fang X, Dou J, Dong X, Duan W (2020) Ferroelectrics 563:31
5. Luo Q, Lu J, Xu X, Jiang F (2017) Ceram Int 43:16178
6. Bakhtiyari AN, Wang K, Wang C, Wu Y (2023) Optik (Stuttg) 290:171333
7. Wen Q, Chen J, Wei X, Lu J, Huang H, Cui C (2022) Ceram Int 48:18842
8. Bakhtiyari AN, Yadav A, Wu Y, Zheng H (2024) Mater Lett 355:135506
9. Wang Q, Zhang Q, Zhang Z, Wang W, Xu J (2020) Ceram Int 46:21461
10. Amina, Ji L, Yan T, Wang Y, Li L (2019) Opt Laser Technol 116:232
11. Xie XZ, Pan ZY, Wei X, Huang FM, Hu W, Hong MH (2011) J Laser Micro/Nanoeng 6:209
12. Bakhtiyari AN, Wu Y, Wang L, Wang Z (2023) J Mater Res Technol 24:663
13. Bakhtiyari AN, Wang Z, Zheng H (2021) J Manuf Process 65:471
14. Bakhtiyari AN, Wang Z, Wang L, Zheng H (2021) Opt Laser Technol 135:106721
15. Bakhtiyari AN, Wu Y, Qi D, Zheng H (2023) Optik (Stuttg) 272:170297
16. Journal AI, Fang X, Wang S, Li B (2022) Integr Ferroelectr 230:179
17. Bakhtiyari AN, Omidi M, Yadav A, Wu Y, Zheng H (2024) Appl Phys A 130:101
18. Ghazvini M, Varedi-koulaei SM, Hossein M, Kim M (2022) Eng Anal Bound Elem 145:363
19. Shahriari B, Swersky K, Wang Z, Adams RP, De Freitas N (2016) Proc IEEE 104:148
20. Yao X, Fu J, Xu Y, He Y (2013) Procedia CIRP 10:91
21. Qin P, Hu H, Yang Z (2021) Sci Rep 1
22. Meraihi Y, Gabis AB (2021) 50001
23. Sultan B, Nantiwat Y, Sujin P, Ali B, Yildiz R, Sait SM (2021) Eng Comput
24. Zahra S, Seyedali M, Shahrzad M, Hossam S, Aljarah I (2017)

Parameter Identification for Constitutive Models via Tree-Structured Parzen Estimator

Supattra Singnisai and Wachirawit Ponghiran

Abstract Finite element method (FEM) is a widely adopted technique for simulating metal foaming processes. Constitutive models, which describe the flow behavior of materials, are critical for FEM accuracy. However, selecting appropriate parameters for these models is challenging, often requiring extensive trial and error. This work explores the use of the Tree-Structured Parzen Estimator (TPE) as an approach to automate the parameter identification process for constitutive models, particularly the Johnson–Cook and Zerilli-Armstrong models. By applying TPE to experimental data from two distinct materials, we demonstrate that TPE significantly improves sample efficiency compared to traditional genetic algorithm methods, reducing the computational requirement to less than 1% while achieving high accuracy of flow stress predictions. This work highlights the use of statistical methods to streamline the development of accurate constitutive models in FEM simulations.

Keywords Parameter identification · Constitutive models · Tree-structured Parzen estimator · Johnson–cook model · Zerilli-Armstrong model

1 Introduction

In recent years, the finite element method (FEM) has emerged as a popular and effective approach for analyzing the metal foaming process and optimizing manufacturing parameters [1, 2]. Constitutive models, which represent the flow behavior of materials, are commonly employed by FEM to simulate material deformation under various loading conditions [3–5]. Hence, the accuracy of these models is crucial as it dictates the quality of simulation results and indicates how closely they match actual physical changes [6, 7]. However, developing an accurate constitutive equation for

---

S. Singnisai
Rajamangala University of Technology Isan, Nakhon Ratchasima, Thailand
e-mail: supattra.si@rmuti.ac.th

W. Ponghiran (✉)
IBM Research Yorktown Height, Yorktown Heights, NY, USA
e-mail: wponghiran@ibm.com

S. Y. Liang (ed.), *Manufacturing Technologies*, Lecture Notes in Mechanical Engineering, https://doi.org/10.1007/978-3-031-95455-9_4

a new material typically requires multiple iterations of selecting and testing model parameters against empirical data. This challenge arises from various factors such as a variation in experimental setup and non-ideal flow behavior of alloy that may deviate from general assumptions of the constitutive models. Therefore, engineers at times come up with different, yet similar, models to describe the same material like AISI 316L [8].

The Johnson–Cook (JC) model is one example of the constitutive models widely used in commercial FEM software to describe the deformation behavior of alloys [8, 9]. Its simplicity and small number of parameters allow for quick calculations of flow stress [6, 7]. However, before the use of this constitutive equation for numerical simulation, it must be calibrated using experimental data. The model equation is typically rearranged to reduce the number of unknown parameters, enabling some values to be determined through linear regression. Ideally, the empirical data should closely follow the JC model, yielding a single value for each parameter. In practice, selecting suitable parameters is often ambiguous, requiring engineers to arbitrarily choose a set of possible values and iteratively test the model to find the best fit (as illustrated in Sect. 2.2). The limitations of simple JC in capturing certain physical phenomena, such as lattice resistance, necessitate the use of alternative constitutive models like the Zerilli-Armstrong (ZA) model [10]. This adds complexity to the development of a precise constitutive model for FEM. To determine the most suitable model for a new material, engineers must evaluate all possible constitutive equations and manually fit the experimental data based on arbitrary parameter choices, leading to a time-consuming process that is difficult to replicate.

Recent work has introduced a method to automatically determine parameters for constitutive equations using genetic algorithms (GA) [11]. This optimization framework, inspired by biological evolution, generates a large population of potential parameters, which are selected and refined through an iterative process. The best-fitting parameters are retained and recombined to create a new set of model parameters over multiple iterations. While this technique enables automated parameter identification, it relies extensively on trial and error. For instance, 500,000 constitutive model fittings were required to identify parameters for the JC model of DH-36 steel [11]. This is because the GA calls for a large set of samples to avoid premature convergence and maintain good-quality solutions in each iteration, resulting in high computation requirements and low sample efficiency [12]. Although other statistical methods could potentially outperform GA in parameter identification efficiency, only the hierarchical modeling optimization technique has been explored in the literature, and its application has been limited to a single simple constitutive model [13].

This work aims to evaluate the effectiveness of the widely adapted Tree-Structured Parzen Estimator (TPE) for automated parameter identification. TPE has been proven to be a flexible method for exploring non-convex and discontinuous search spaces, leading to state-of-the-art models in various domains [14]. In this study, we demonstrate that TPE can be effectively adapted to identify parameters for both JC and ZA models on experimental data from two different materials: Ti-6Al-4V (Ti64) titanium alloy and DH-36 steel. Our results show that the TPE method is significantly more sample-efficient than the GA-based approaches, requiring only 2% of the model fittings to achieve similar and better quality in terms of predicted flow stress. Consequently, our method provides a faster and less computationally intensive approach to parameter identification. The source code for reproducing all experiments will be made publicly available upon acceptance.

2 Background

2.1 Constitutive Models to Predict Flow Stress

The Johnson–Cook model and the Zerilli-Armstrong model are two constitutive models that are frequently used for FEM simulations to describe the relationship between flow stress and strain of materials under a wide range of strain rates and temperature due to their simple forms [6, 7]. We describe each of them in detail in the following subsections.

The Johnson–Cook (JC) model captures the relationship between flow stress and strain by accounting for three important physical phenomena: isotropic hardening, strain rate hardening, and thermal softening [9]. Isotropic hardening describes the expansion of the yield surface without any distortion or translation while strain hardening refers to the material strengthening as it deforms. Thermal softening, on the other hand, explains how materials weaken at elevated temperatures. The JC model combines these effects by treating them as independent factors, allowing their combined influence on flow stress to be expressed through three separate terms as follows:

$$\sigma_y = \left(A + B\varepsilon_p^n\right)\left(1 + C\ln\left(\dot{\varepsilon}_p^*\right)\right)\left(1 - T^{*m}\right) \tag{1}$$

where σ_y represents the flow stress and ε_p represents the true strain. A, B, C, m, and n are material-specific parameters related to the aforementioned phenomena. $\dot{\varepsilon}_p^* = \dot{\varepsilon}_p/\dot{\varepsilon}_0$ is a dimensionless strain rate normalized by a reference value while $T^* = (T - T_0)/(T_m - T_0)$ is homologous temperature where T is the temperature, T_0 is the reference temperature and T_m is the melting temperature. Although the JC model effectively predicts flow stress for various materials [10], it has limitations, particularly in the lack of explicit consideration of the underlying dislocation mechanics that govern deformation. The JC model does not distinguish between athermal and thermally activated dislocation processes; hence it may lead to non-physical temperature scaling of the predicted flow stress.

The Zerilli-Armstrong (ZA) model was developed to address the limitations of the JC model, offering a more physically grounded approach to flow stress prediction, especially for specific alloy types like Ti64 [11, 15]. The ZA model comes in several variants, each tailored to capture the behavior of material with different crystal structures. The body-centered cubic (BCC) variant, for example, focuses on dislocation behavior in BCC alloys, where dislocation motion is governed by Peierls–Nabarro stress (lattice resistance), leading to limited strain hardening and strain rate-independent work hardening [16]. This variant can be expressed mathematically as:

$$\sigma_y = \sigma_0 + K\varepsilon_p^n + Be^{-(\beta_0-\beta_1\ln(\dot{\varepsilon}_p^*))T} \tag{2}$$

where σ_0, K, n, B, β_0, and β_1 are material-specific parameters. ε_p, $\dot{\varepsilon}_p^*$, and T are true strain, normalized strain rate, and temperature similar to the JC model.

Another variant, the face-centered cubic (FCC) model, is designed for FCC alloys, where dislocation interactions is dominant, resulting in significant strain hardening. This model variant can be mathematically expressed as:

$$\sigma_y = \sigma_0 + K\varepsilon_p^n + B_0\sqrt{\varepsilon_p}e^{-(\alpha_0-\alpha_1\ln(\dot{\varepsilon}_p^*))T} \tag{3}$$

where B_0, α_0, and α_1 are new material-specific parameters.

Nonetheless, not all alloys conform to the characteristics of FCC or BCC metals, so a combined ZA model has been utilized to capture the behavior of materials with different underlying crystal structures like hexagonal close-packed structure. The expression for the general ZA model is as follows:

$$\sigma_y = \sigma_0 + K\varepsilon_p^n + B_0\sqrt{\varepsilon_p}e^{-(\alpha_0-\alpha_1\ln(\dot{\varepsilon}_p^*))T} + Be^{-(\beta_0-\beta_1\ln(\dot{\varepsilon}_p^*))T} \tag{4}$$

2.2 *Method to Manually Determine Model Parameters and Its Problem*

Before using any constitutive equation in a numerical simulation, the model must first be calibrated using empirical data. For the Johnson–Cook (JC) model, this calibration involves determining five model parameters: A, B, C, m, and n. By analyzing experimental results at the reference temperature and strain rate, Eq. (1) can be simplified to a linear expression of A, B and n as:

$$\ln(\sigma_y - A) = \ln(B) + n\ln(\varepsilon_p). \tag{5}$$

The parameters B and n can then be determined by plotting the logarithm of flow stress against the logarithm of strain and fitting the data to a linear equation as $y = b + ax = \ln(B) + nx$. Once B and n are identified, the parameter C can be determined by simplifying the JC equation at the reference temperature, reducing it to:

$$\sigma_y/\left(A + B\varepsilon_p^n\right) = 1 + C\ln\left(\dot{\varepsilon}_p^*\right) \tag{6}$$

Here, C corresponds to the slope of the linear regression line plotted between $\sigma_y/\left(A + B\varepsilon_p^n\right)$ and $\ln(\dot{\varepsilon}_p)$. A similar process is used to determine the value of m by considering the JC equation at the reference strain rate, simplifying it to:

$$\ln\left(1 - \sigma_y/\left(A + B\varepsilon_p^n\right)\right) = m\ln(T) \tag{7}$$

The parameter m is then derived as the slope of the best-fitting linear equation for the data points at the reference strain rate.

The issue of manual parameter determination is highlighted in Fig. 1, which illustrates the challenges in identifying the parameters of the JC model for Ti64 steel [15]. Engineers often encounter variability in identifying parameters because the experimental results do not perfectly align with the linear assumptions of the model. As a result, the plotted data do not form a straight line. On the left side of Fig. 1, linear Eqs. 1 and 2 provide an option that generally fits the overall data, but with noticeable errors at strain rates of e^{-1} and beyond. On the other hand, linear Eq. 3 offers an almost perfect fit for data points at strain rates between e^{-3} and e^{-1} but have larger error for outside the range. The choice between these options is unclear and arbitrary to engineers who perform model calibration. Furthermore, since the parameter identification is a sequential process, the choice of parameters B and n influences the selection of parameter C on the right of Fig. 1. Multiple options exist for obtaining parameter C through linear regression, and the best choice among these options is not obvious. Both linear Eqs. 4 and 5 may result in sufficiently good constitutive models that match the experimental data. However, it is difficult to justify the selection until all five parameters are determined and the complete equation is evaluated against empirical data. If the chosen parameters result in a poor fit, engineers must restart the manual selection process with different choices.

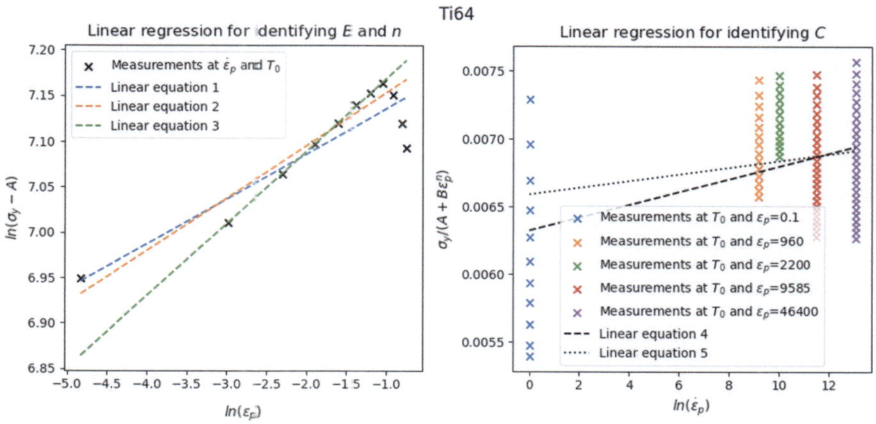

Fig. 1 Plots and linear questions used for identifying parameters of the JC model: parameter B and n (left) and parameter C (right). Experiment data of Ti64 is taken from [15]

3 Methods

3.1 *Tree-Structured Parzen Estimator for Parameter Identification*

The Tree-Structured Parzen Estimator (TPE) is a Bayesian optimization method that has been widely adopted in many recent frameworks, such as parameter search for a drug discovery [17] and financial modeling [18]. The core problem in these applications is to find a set of model parameters $\boldsymbol{x}$ that minimizes an objective function $f(\boldsymbol{x})$ such that:

$$\boldsymbol{x}_{opt} \in \operatorname{argmin} f(\boldsymbol{x}) \text{ given } \boldsymbol{x} \in X. \tag{8}$$

In the context of this paper, the problem of identifying parameters for the constitutive model can be framed similarly, where the goal is to minimize the difference between empirical data and model predictions. The technique of minimizing this difference by guessing one parameter at a time is one approach to satisfy the Eq. (5). However, TPE can achieve this goal in a more systematic way and without requiring engineer intervention. TPE iteratively searches for the optimal parameters $\boldsymbol{x}_{opt}$ using an acquisition function to guide between exploration (searching unexplored regions) and exploitation (refining around promising observations). One of the common choices for the acquisition function is Expected Improvement (EI), a statistical term that can be computed easily through an observation of objective function $\boldsymbol{y} = f(\boldsymbol{x})$

$$EI[\boldsymbol{x}|D] = \int_{-\infty}^{y^*} (y - y)p(y|\boldsymbol{x}, D)dy \tag{9}$$

where $p(y|\boldsymbol{x}, D)$ is the probability density function (PDF) based on a set of observations D. The core idea behind TPE lies in how it estimates the density function p. TPE works by first sampling a set of model parameters randomly to evaluate the objective function. It then simulates the parameter space using two PDFs: one for configurations that yield better (lower) objective values and another for those yielding worse (higher) values. TPE then selects new parameters to evaluate by maximizing the ratio of these two PDFs, focusing on regions in the parameter space that are more likely to yield better results.

Compared to the existing work that uses genetic algorithms, which rely on stochastic operators like mutation and crossover to explore a parameter space, TPE is more efficient in focusing the search on promising regions using a statistical mean. Genetic algorithms in general require many evaluations to converge to a good solution due to their reliance on random trial and error. In contrast, TPE leverages probabilistic modeling, as shown in Eq. (6), to make more informed decisions and reduce the number of required model fittings, leading to faster convergence with fewer computational resources. This makes TPE particularly advantageous in scenarios when the search space is large and complex, such as in model identification for a new material. Throughout this work, we utilize an off-the-shelf TPE package available public only from [21] and evaluate it against existing work from [11].

3.2 *Experimental Data for Constitutive Model*

To demonstrate the application of TPE, we utilize experimental results from existing literature that model the flow stress behavior of metal and alloys. Data for Ti-6Al-4V (Ti64) titanium alloy are from [15] while the data for DH-36 is from [19, 20]. Both are the same experimental data that we used in the previous work. For a fair comparison with the existing work, only experimental data at room temperature is considered by TPE and GA.

4 Results and Discussion

4.1 *Automated Parameter Identification Results*

Figure 2 shows the flow stress prediction results for the Ti64 titanium alloy at different strain rates using the JC model and three variants of the ZA model. All parameters were identified using the proposed TPE method without requiring engineer intervention to demonstrate its effectiveness in allowing engineers to quickly visualize and select a constitutive model that accurately represents the empirical data for use in FEM simulations. We can easily observe from the results that the JC model consistently follows the experimental trends across all strain rates, though it tends to overestimate flow stress at higher strain rates, particularly at $\varepsilon_p = 9585$. In contrast, the ZA BCC model shows better alignment with experimental data at both high and low strain rates, closely predicting the observed stress behavior. While other ZA model variants perform worse, the ZA general model stands out at high strain rates, capturing the flow stress trend more accurately than the others. Overall, this indicates that engineers can use our method to identify constitutive model parameters and select the ZA BCC model with confidence for FEM simulations of Ti64 at low strain rates.

Similarly, we demonstrate the utilization of the proposed method to identify parameters of the JC and ZA model for DH36 steel as shown in Fig. 3. In this experiment, all models align relatively well with the empirical data, particularly the JC model, which supports previous findings in the literature that DH36 steel's flow stress behavior is well-captured by the JC model [19, 20]. The ZA general model tends to slightly overestimate the flow stress under both strain rate conditions while

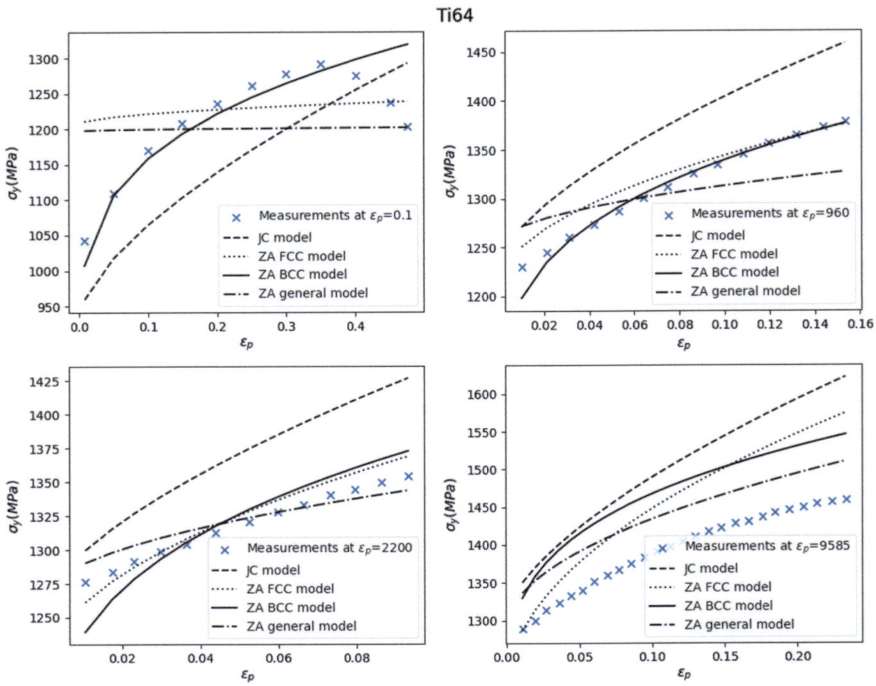

Fig. 2 Flow stress prediction for Ti64 titanium alloy using JC and variant of ZA models at different strain rates: $\varepsilon_p = 0.1$ (upper left), $\varepsilon_p = 960$ (upper right), $\varepsilon_p = 2200$ (lower left), and $\varepsilon_p = 9585$ (lower right) where all parameters are identified with the proposed TPE method

the ZA BCC model tends to slightly underestimate it. Overall, these results demonstrate that the proposed method for parameter identification with TPE is a versatile framework that can be applied to various materials without requiring modifications.

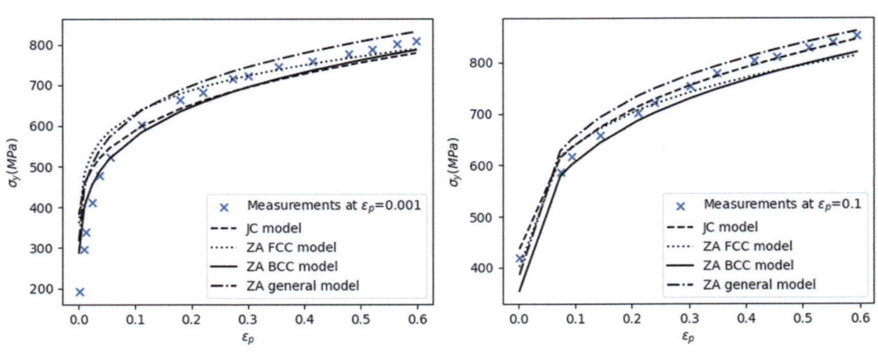

Fig. 3 Flow stress prediction for DH36 steel using JC and variant of ZA models at different strain rates: $\varepsilon_p = 0.001$ (left) and $\varepsilon_p = 0.1$ (right) where all model parameters are identified with the proposed TPE method

4.2 Effectiveness of TPE in Parameter Identification

To evaluate the effectiveness of the TPE in parameter identification compared to the GA approach proposed in previous work [11], we compared the Root Mean Square (RMS) error between empirical flow stress data and model predictions. The model predictions were generated using the best set of parameters identified at specific points during the parameter identification process.

Figure 4 illustrates the RMS error reduction when using TPE and GA to identify constitutive model parameters for the Ti64 titanium alloy. In terms of the parameter identification performance, the TPE achieves lower RMS errors across most models compared to GA runs except the JC one, in which GA discovers a slightly better set of model parameters. This shows that TPE is a reliable technique to come up with an accurate constitutive model. In terms of parameter identification efficiency, the TPE method reaches a lower RMS error with fewer model fittings, while the GA often requires more iterations to achieve comparable results. For instance, in the JC model, TPE rapidly reduces the RMS error, reaching a lower value within 1,000 trials of model fitting. This is in contrast to the GA, which typically relies on a large number of trial-and-error iterations. The efficiency of TPE becomes particularly evident in the variants of the ZA models.

It is important to note that TPE may not always outperform the genetic algorithm in both quality and speed. Since GA involves extensive trial and error, it is possible for GA to run into a model parameter with slightly lower RMS errors, as seen in the JC model results at the end of the 3,000-model fitting. However, overall performance and speed are critical criteria for an effective parameter identification technique, and TPE demonstrates strong advantages in these areas. For instance, the existing work reports that GA required around 500,000 computations to reach a local minimum, whereas TPE achieved similar results with less than 1% of that, requiring only 3,000 model fittings.

Similarly, Fig. 5 compares the RMS error reduction when using TPE and GA to identify constitutive model parameters for DH36 steel. A rapid reduction in RMS error with fewer model fitting trials is observed with TPE. We also note that the rate of RMS error reduction appears to correlate with the complexity of the search space. For DH36 steel, parameter identification appears to be easier and RMS error reduction is relatively faster than the experiment with Ti64 titanium alloy. This could be a useful indication that certain models may be inherently more suitable for specific materials than others and will be further investigated in future study.

Throughout all experiments, we observed that searching for parameters in logarithmic space for some parameters leads to faster convergence, like m and n of the JC model and parameter alpha and beta of the ZA. We hypothesize that this is because these parameters are exponent terms in the constitutive equations. Scaling them in logarithmic space allows for more effective adjustments, especially when the exponent values are small or close to zero.

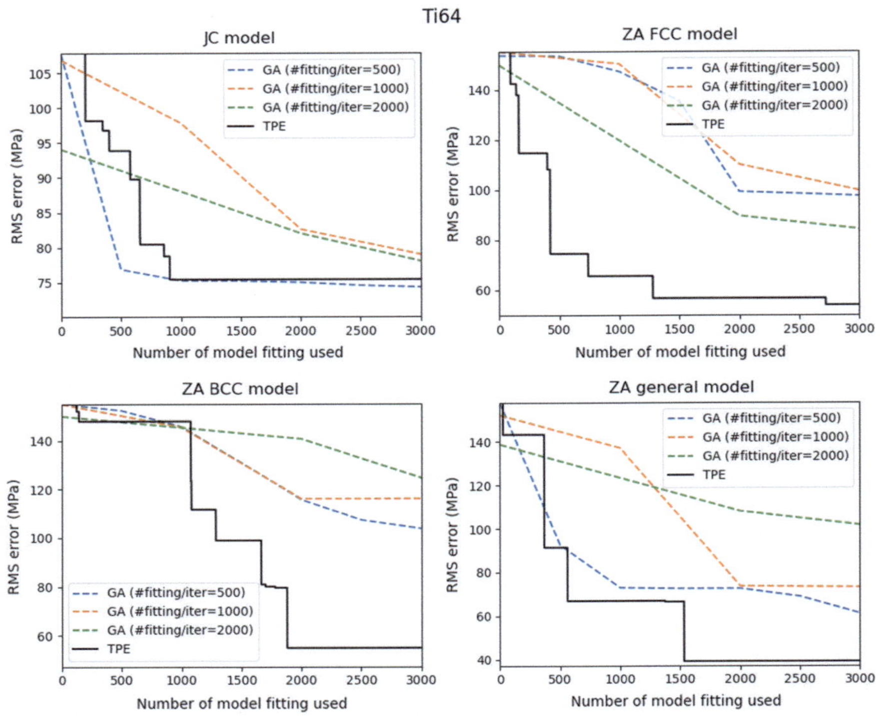

Fig. 4 Comparison of RMS error reduction when using TPE and GA to identify constitutive model parameters for Ti64 titanium alloy. Four different constitutive models are used as means for comparing the sample efficiency of TPE: JC model (upper left), ZA FCC model (upper right, ZA BCC model (lower left), and ZA general model (lower right)

5 Conclusion

This work presents the TPE as a robust and efficient method for automated parameter identification of the Johnson–Cook and Zerilli-Armstrong models that are widely used in FEM. Through experiments on Ti-6Al-4V titanium alloy and DH-36 steel datasets, we demonstrate that TPE achieves accuracy comparable to the existing GA technique. Analysis of error rate progression during model fitting reveals that TPE efficiently navigates the parameter search space, requiring only 1% of the trial-and-error iterations needed by the model fitting through the GA technique. This result underscores the value of statistical methods in improving the accuracy of metal deformation predictions, making it a valuable tool for FEM on material modeling and manufacturing optimization. Since the proposed method is model agnostic, it can be applied to other constitutive models as well. Future work will involve evaluating the proposed approach against the GA-driven parameter search technique using data from commonly used metal alloys such as Weldox 700 [22] and 304 Stainless Steel [23]. We also plan to extend the method to fit parameters of other constitutive models

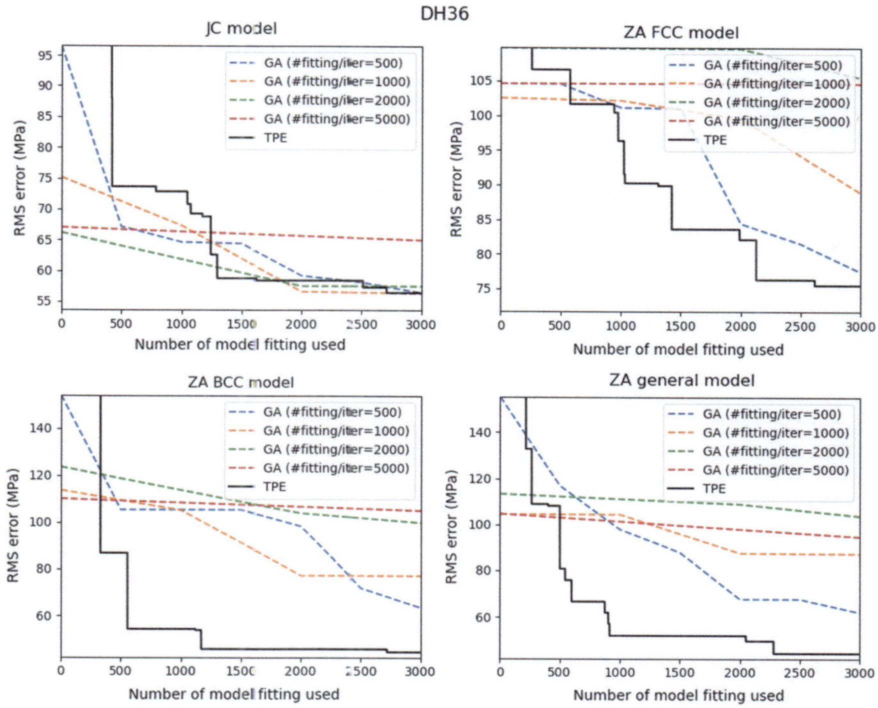

Fig. 5 Comparison of RMS error reduction when using TPE and GA to identify constitutive model parameters for DH36 steel. Four different constitutive models are used as means for comparing the sample efficiency of TPE: JC model (upper left), ZA FCC model (upper right, ZA BCC model (lower left), and ZA general model (lower right)

for materials like concrete, that is frequently used in finite element analysis of large structures [24]. This approach is expected to significantly reduce the manual effort involving in the constitutive model parameter identification.

References

1. Li H-Y et al (2011) Constitutive equation to predict elevated temperature flow stress of V150 grade oil casing steel. Mater Sci Eng A 530:367–372
2. Changizian P, Zarei-Hanzaki A, Roostaei AA (2012) The high temperature flow behavior modeling of AZ81 magnesium alloy considering strain effects. Mater Des 39:384–389
3. Li C et al (2010) Constitutive equations in creep of 7B04 aluminum alloys. Mater Sci Eng A 527(16–17):3623–3629
4. Han Y et al (2013) A comparative study on constitutive relationship of as-cast 904L austenitic stainless steel during hot deformation based on Arrhenius-type and artificial neural network models. Comput Mater Sci 67:93–103
5. Samantaray D, Mandal S, Bhaduri AK (2010) Constitutive analysis to predict high-temperature flow stress in modified 9Cr–1Mo (P91) steel. Mater Des 31(2):981–984

6. Xiao Y-H, Guo C (2011) Constitutive modelling for high temperature behavior of 1Cr12Ni3Mo2VNbN martensitic steel. Mater Sci Eng A 528(15):5081–5087
7. Ji G et al (2011) A comparative study on Arrhenius-type constitutive model and artificial neural network model to predict high-temperature deformation behaviour in Aermet100 steel. Mater Sci Eng A 528(13–14):4774–4782
8. Umbrello D, M'saoubi R, Outeiro JC (2007) The influence of Johnson–Cook material constants on finite element simulation of machining of AISI 316L steel. Int J Mach Tools Manuf 47(3–4):462–470
9. Johnson GR (1983) A constitutive model and data for metals subjected to large strains, high strain rates and high temperatures. In: Proceedings of the 7th international symposium on ballistics, The Hague, Netherlands
10. Armstrong RW, Zerilli FJ (1994) Dislocation mechanics aspects of plastic instability and shear banding. Mech Mater 17(2–3):319–327
11. Robson JD et al (2024) Calibration of constitutive models using genetic algorithms. Mech Mater 189:104881
12. Mitchell M (1998) An introduction to genetic algorithms. MIT Press
13. Papadimas N, Dodwell. T (2021) A hierarchical Bayesian approach for calibration of stochastic material models. Data-Centric Eng 2:e20
14. Watanabe S (2023) Tree-structured Parzen estimator: understanding its algorithm components and their roles for better empirical performance. arXiv:2304.11127
15. Meyer HW (2006) A modified Zerilli-Armstrong constitutive model describing the strength and localizing behavior of Ti-6Al-4V. Army Research Laboratory
16. Sim KH et al (2023) Modified Zerilli-Armstrong and Khan-Huang-Liang constitutive models to predict hot deformation behavior in a powder metallurgy Ti-22Al-25Nb alloy. Vacuum 210:111749
17. Schneider P et al (2020) Rethinking drug design in the artificial intelligence era. Nat Rev Drug Disc 19(5):353–364
18. Jafar SH (2022) Financial applications of Gaussian processes and Bayesian optimization. In: Bayesian reasoning and Gaussian processes for machine learning applications. Chapman and Hall/CRC, pp 111–122
19. Nemat-Nasser S, Guo W-G (2003) Thermomechanical response of DH-36 structural steel over a wide range of strain rates and temperatures. Mech Mater 35(11):1023–1047
20. Klepaczko JR et al (2009) Modelling of thermo-viscoplastic behaviour of DH-36 and Weldox 460-E structural steels at wide ranges of strain rates and temperatures, comparison of constitutive relations for impact problems. Mech Mater 41(5):599–621
21. Bergstra J et al (2015) Hyperopt: a python library for model selection and hyperparameter optimization. Comput Sci Disc 8(1):014008
22. Børvik T et al (2001) A computational model of viscoplasticity and ductile damage for impact and penetration. Eur J Mech A/Solids 20(5):685–712
23. Jia B et al (2021) Simple shear behavior and constitutive modeling of 304 stainless steel over a wide range of strain rates and temperatures. Int J Impact Eng 154:103896
24. Sima JF, Roca P, Molins C (2008) Cyclic constitutive model for concrete. Eng Struct 30(3):695–706

Digital Manufacturing System and Control Technology

Straightness Error Measurement System of Machine Tools Base on Eddy Current Displacement Sensor

Jung-Tang Huang and Yu-Cian Yao

Abstract Various errors in machine tools can significantly impact the final quality of machining, with geometric errors being the primary source of such influence. Geometric errors comprise a combination of multiple factors, including angular, positioning, straightness, and perpendicularity errors across the machine's axes, amounting to 21 different types. Typically, correcting these errors requires the use of a laser interferometer. However, the complex operation of laser interferometers leads to lengthy setup times, and the high cost of the laser equipment, including the necessary mirrors for different error measurements, further complicates the process. Additionally, the precision required by laser measurements necessitates a clean and stable environment, often resulting in considerable time loss during measurement. Consequently, many opt to extend the calibration intervals to reduce costs, which makes it difficult to capture dynamic geometric changes in the machine during operation. Failure to promptly assess the machine's health can lead to suboptimal machining quality and precision, increasing production costs. This study proposes a straightness error measurement system designed to significantly reduce the time cost associated with error measurement. The system involves installing an eddy current micro-displacement sensor, with an accuracy of 1.3 μm and a resolution of 200 nm, into an orthogonal measurement device mounted on the tool-holder. Using a ball array as the workpiece, the system measures the errors between each measurement point in the array, calculating the machine's horizontal and vertical straightness. The obtained error values are then input into the CNC controller for compensation. This method reduces the reliance on high-precision instruments and minimizes downtime associated with maintenance, thereby lowering overall time costs.

Keywords Machining accuracy · Geometric error · Straightness error · Eddy current displacement sensor

---

J.-T. Huang (✉) · Y.-C. Yao
National Taipei University of Technology, Taipei, Taiwan
e-mail: jthuang@ntut.edu.tw

S. Y. Liang (ed.), *Manufacturing Technologies*, Lecture Notes in Mechanical Engineering, https://doi.org/10.1007/978-3-031-95455-9_5

1 Introduction

This study proposes a machine tool geometric error compensation system that can measure machine tool straightness errors. Using an eddy current displacement sensor with a resolution of 200 nm, and measuring the straightness error at three-dimensional measurement points on the workpiece, the straightness error on each transmission shaft is measured, and finally the error compensation value is input to the CNC control device to compensate. The purpose is to reduce the accuracy of the instrument and frequency of use while reducing maintenance downtime.

2 System Structure

A multi-body system (MBS) developed by J. W. Fan et al. was based on rigid body kinematics, in which positioning error and motion error were incorporated, proposing a general kinematic model for constructing CNC machine tools. The machine tool structure was modeled by them as a composition of several rigid bodies [1]. To ensure the machining quality of the machine tool, regular calibration is necessary to maintain its accuracy. After extended cutting operation, significant wear occurs on the contact surface between the slide rail and the guide groove of the machine tool, thus reducing the accuracy of the machine table [2]. The system architecture diagram shown in Fig. 1 represents a geometric error measurement framework developed based on eddy current sensors and the VMX (Virtual Multi-aXis) smart box. This framework includes our previously developed angular error measurement system [3] and a straightness measurement system developed in this research. The straightness measurement tool is mounted on the machine tool, and a Bluetooth receiver is used to collect the measured data and integrate the displacement changes in the angular error measurement system utilizing eddy current sensors installed inside the machine tool to measure the raw displacement data of various measurement points along the guide rails. After collecting both the angular error and straightness error data, the data is mapped to the machine tool's mechanical coordinates and preprocessed. Once the data's reliability has been verified against laser interferometer measurements, the geometric errors detected by the eddy current sensors are transformed into overall machine tool geometric error values using homogeneous transformation matrices. These values are then input into the CNC controller as compensation parameters, completing the geometric error compensation of the entire machine tool.

The VMX as shown in Fig. 1, developed by the Tool Machine Center of the Industrial Technology Research Institute in Taiwan, is a cross-controller communication module that can capture coordinate and status data from various brands of machine tools, as well as sensor data. It enables data merging, meaning that VMX can integrate information from CNC controllers and measurement data from machines and display it in a software interface. This includes a CNC communication interface, data acquisition interface, database, and communication functions between nodes.

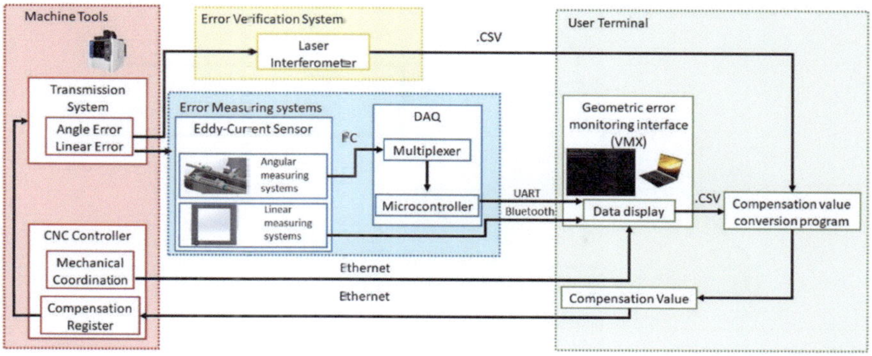

Fig. 1 Geometric error measurement system architecture diagram

Once the VMX platform is loaded, apps with these functional modules can be used, and it also allows developers to create multifunctional apps through multiple APIs, lowering the development threshold for industrial software in machine tools.

3 Principals of Eddy Current Sensor

The operation of the eddy current sensor is based on RLC resonance. If an alternating current is passed through an inductive coil, an alternating magnetic field will be generated by the coil. According to the law of electromagnetic induction, when the alternating magnetic field comes close to a conductor, an induced magnetic field and eddy currents in the opposite direction to the coil's alternating magnetic field will be generated on the surface of the conductor. Each weakens the other, these two magnetic fields. As shown in Fig. 2, when the distance between the coil and the conductor changes, resonance is caused by the interaction between the magnetic fields. Therefore, by measuring the coil impedance, the displacement change between the coil and the conductor can be determined, as the coil inductance value of the circuit changes [4].

4 Experimental Method

Most influential among the many factors that affect the machining accuracy during the machining process is the geometric error caused by the components and structure of the bed. Composed of 21 geometric error components in total is the volume error in a three-axis machine tool, which includes three perpendicularity errors, one positioning error, six straightness errors, and nine angle errors [5].

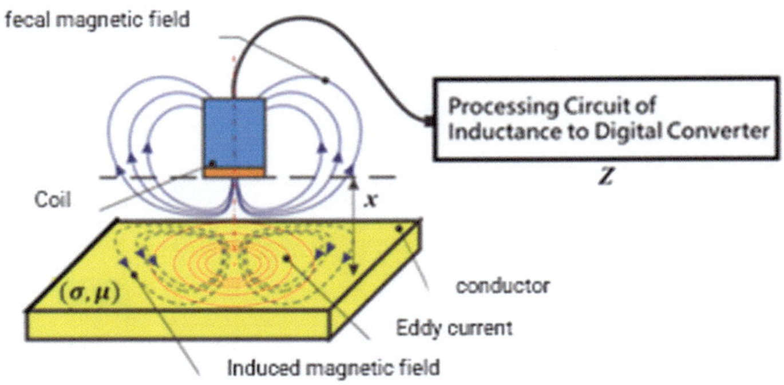

Fig. 2 Eddy current sensor

4.1 Straightness Error Measurement Workpiece

In this project, a custom artifact was fabricated, featuring drilled circular holes where low-carbon steel balls are placed for measurement purposes. By utilizing an orthogonal tool-holder type sensor to measure the low-carbon steel balls on the artifact [6], the displacement of each ball's center from its ideal position in the X, Y, and Z directions can be calculated. These displacements are then converted into straightness errors and verticalness errors. The physical depiction of the artifact is shown in Fig. 3.

Fig. 3 3-D straightness measurement artifact

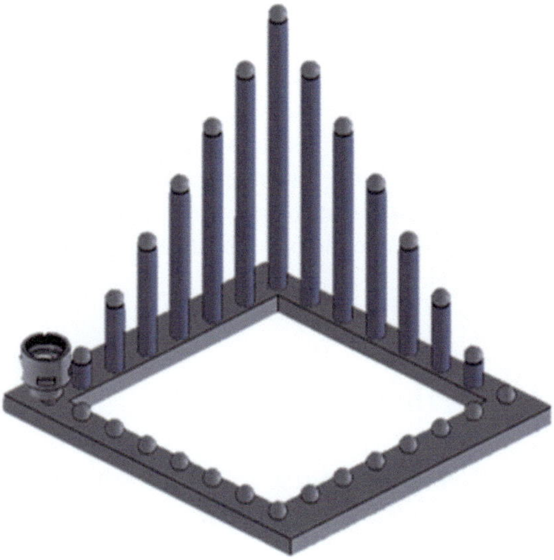

4.2 Tool-Holder Type Eddy Current Sensor Device

In this study, an orthogonal tool-holder sensing device was developed, where an eddy current sensor measures in the X, Y, and Z directions [7]. However, the relationship between the coil impedance Z_{eff} of the eddy current sensor and the distance x between the coil and the conductor can be defined by Eq. (1). In this equation, R_c represents the resistance of the sensor, R_e represents the resistance generated by eddy currents, σ is the electrical conductivity of the conductor, μ is the magnetic permeability of the conductor, f is the operating frequency of the sensor, ω is the angular frequency of excitation, L_c is the inductance of the sensor, and L_e is the inductance generated by eddy currents. From Eq. (1), it is evident that by controlling the inductance and resistance of the sensor's resonant circuit, the material properties of the conductor, the sensor frequency, and the angular frequency of excitation, the displacement distance can be derived from the coil impedance. Since the device is mounted on the spindle, it can be used to measure the three-dimensional straightness of a workpiece by lowering the machine tool's spindle and calculating the error values between the measurement points on the artifact. As the sensor is non-contact, the orthogonal sensor's characteristics are utilized during positioning. By calculating the distance between the known position of the sensor and the measurement points, the measurement location is adjusted so that the measurement points are aligned with the center of both sensors [8]. The tool-holder type eddy current sensing device is shown in the Fig. 4.

$$Z_{\text{eff}} = \left[R_c + R_e(x, \sigma, \mu, f)\right] + j\omega\left[L_c - L_e(x, \sigma, \mu, f)\right] \tag{1}$$

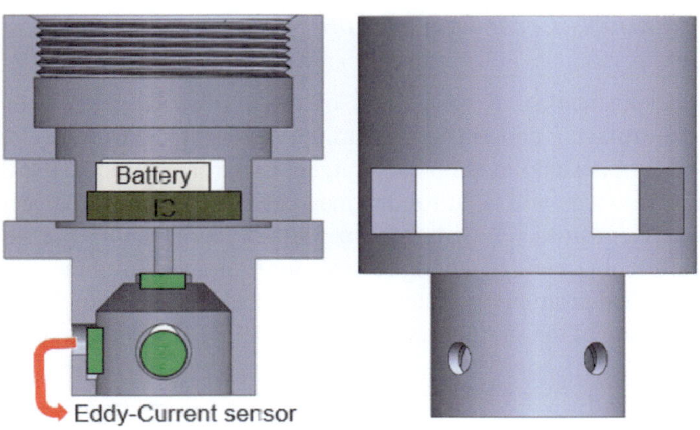

Fig. 4 Tool-holder type eddy current sensor device

4.3 Measurement Standard

Since the laser interferometer machine used is a measuring machine that complies with ISO standards [9], the straightness accuracy measured by the laser interferometer and eddy current joint can be compared using Renishaw's analytical accuracy method, which is mainly used for forward repeatability Rep (+), reverse repeatability Rep (−), repeatability in both directions and accuracy are analyzed. In this study, the horizontal straightness measured by the two methods is analyzed using Renishaw analysis standards for comparison. The rest of the measurement standards are measured according to ISO230-2 [10]; the primary content focuses on testing the repeatability and positioning accuracy of the location measurement system. Repeatability refers to the consistency of results when the machine tool repeatedly measures the same position under identical conditions, which is crucial for assessing the machine's stability and precision. Positioning error, on the other hand, refers to the difference between the machine's actual positioning point and the target positioning point, serving as a key indicator for evaluating the machine's accuracy. The standard provides detailed instructions on how to test and record these parameters, including which testing equipment to use, how to set up the testing environment, and how to analyze and report the test results.

The specific testing methods involve the use of high-precision measurement equipment such as laser interferometers for location measurements and require that tests be conducted under specified environmental conditions to ensure accuracy and consistency of the results. This standard aims to standardize the testing methods and evaluation criteria for machine tool position measurement systems, ensuring the precision and reliability of the machines.

4.4 Straightness Error Calculation

Involving the positioning of the centers of various measurement points on the measurement artifact, a coordinate measuring machine (CMM, Zeiss Contura G2) will be used in this study to calibrate straightness error. Subsequently, the sensor developed in this study will be utilized to measure the spherical test object installed at the measurement artifact. Based on the measured values, the offset of the center of the spherical test object will be computed, revealing the deviation of the measurement point from its ideal position calibrated by CMM in advance. Figure 5 illustrates the schematic diagram of the straightness error presented on the measurement artifact.

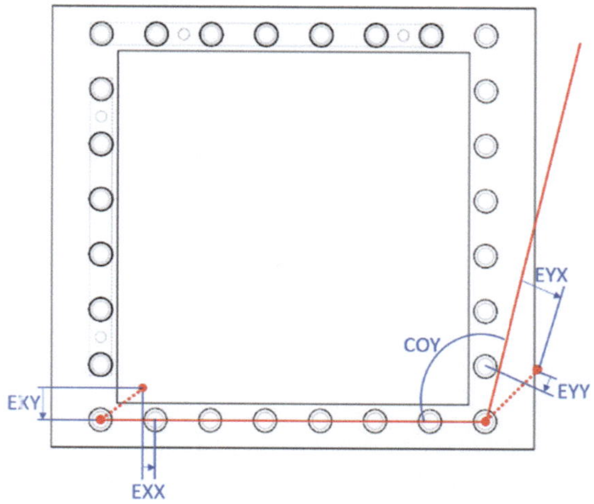

Fig. 5 Straightness error presented on the measurement artifact

4.5 Straightness Error Measurement Process

First the measurement points 1 to 8 in the X-axis direction are measured, from which we obtain the Horizontal Straightness Error ($\delta_y(x)$), Vertical Straightness Error ($\delta_z(x)$), and Linearity Error ($\delta_x(x)$) on the X-axis. Following this, the measurement points 8 to 15 are measured to obtain the straightness errors of the Y-axis, namely $\delta_x(y)$, $\delta_z(y)$, and $\delta_y(y)$. Next, the step measurement points in the same direction as the X-direction (point 15 to point 22) are measured to obtain the Z-axis's $\delta_x(z)$, $\delta_y(z)$, and $\delta_z(z)$. Finally, the accuracy of the Z-axis straightness error is verified by measuring from step measurement point 22 to point 1 in the same direction as the Y-axis, as depicted in Fig. 6.

5 Measurement Result

5.1 Straightness Error Measurement Result

Figures 7 and 8 depict the comparison of average position differences between the straightness measurements obtained using orthogonal tool-holder type eddy current sensors and those from laser interferometry for the X and Y axes, respectively. The horizontal axis represents the machine coordinate positions, while the vertical axis denotes the straightness errors. The blue line represents the measurements from the laser interferometer, and the red line represents the results from the orthogonal tool-holder type eddy current sensors.

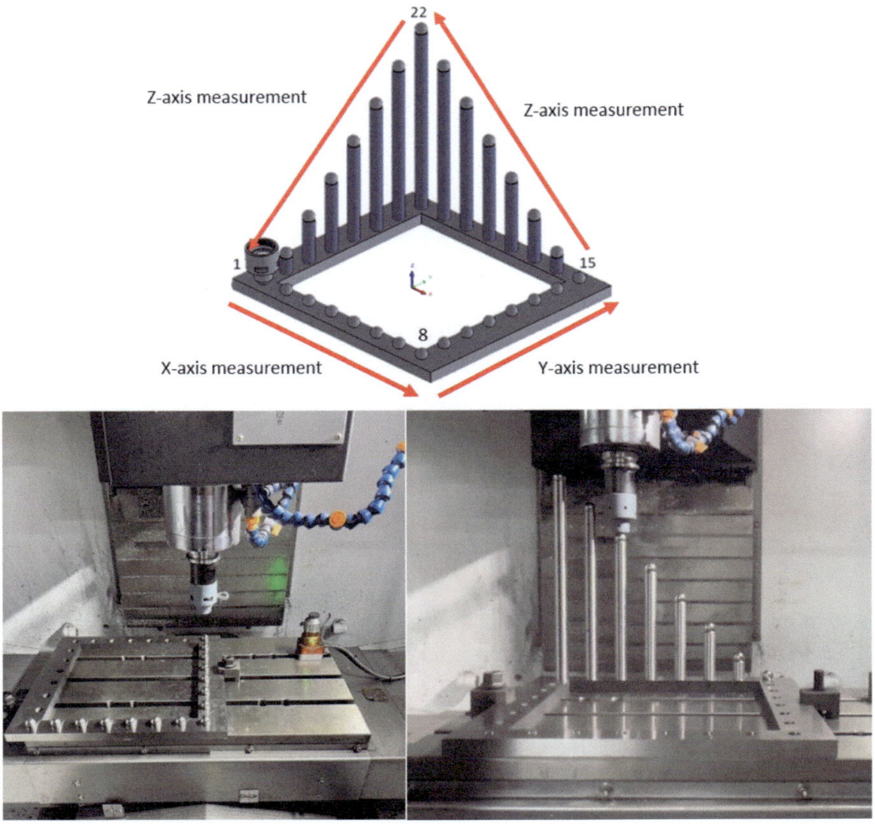

Fig. 6 Straightness error measurement process with tool-holder type eddy current sensors

We analyzed the measurement results and laser interferometry data into Tables 1 and 2. It can be observed that the measurement results used in this study are similar to those of laser interferometer.

5.2 Straightness Error Compensation Result

By integrating these measurements with a previously developed angular error measurement system, we can obtain all the error parameters required for a complete homogeneous equation. We combine the angular error results obtained earlier with the horizontal straightness, vertical straightness, and linearity errors measured in this study, and input these into the homogeneous equation to calculate the overall positioning error of the machine. A comparison of the positioning errors along the X-axis and Y-axis measured by the laser interferometer before and after compensation, wherein the compensation results are shown in Fig. 9.

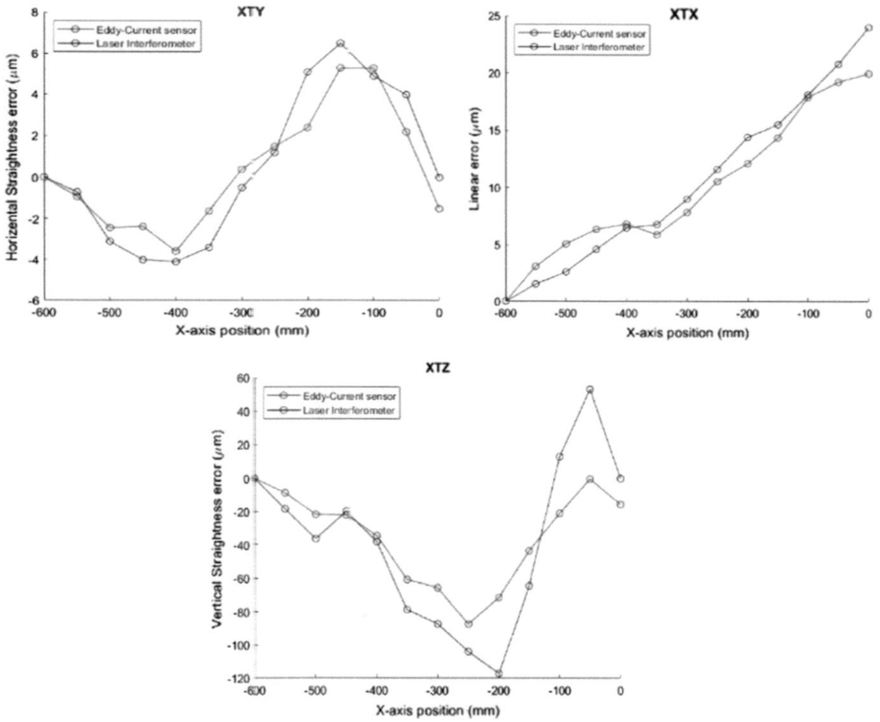

Fig. 7 Comparison of X-axis straightness error measured by two methods

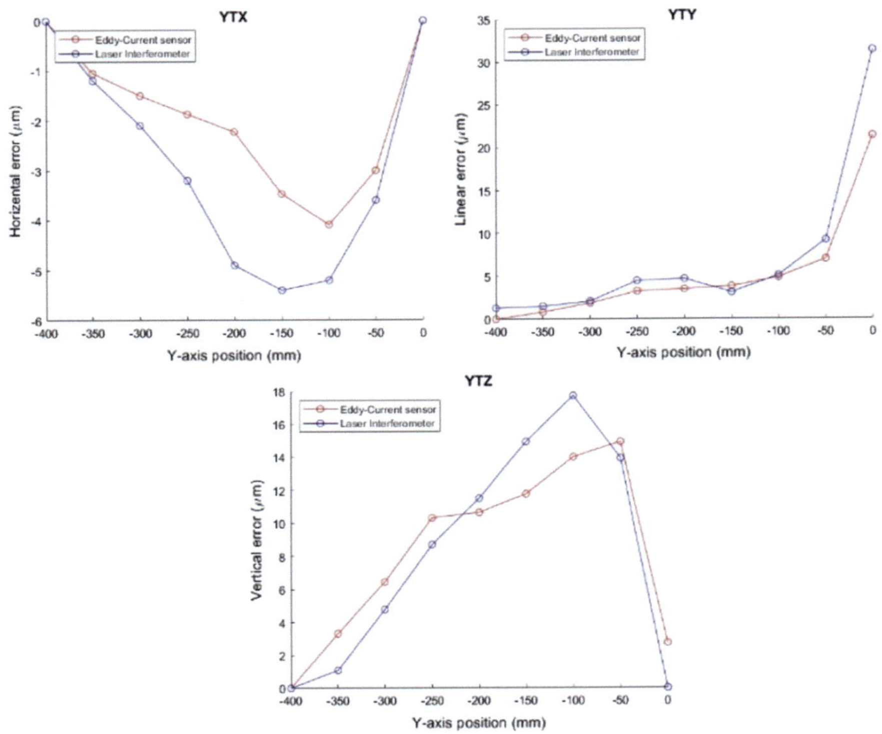

Fig. 8 Comparison of Y-axis straightness error measured by two methods

Table 1 Precision comparison of X-axis error measured by two methods

Unit: um	Laser interferometer			Eddy current sensor		
Error type	$\delta_Y(x)$	$\delta_Z(x)$	$\delta_X(x)$	$\delta_Y(x)$	$\delta_Z(x)$	$\delta_X(x)$
Position accuracy A	11.4	201.7	26.0	10.9	126.5	24.1
Rep(+)	1.7	38.8	4.0	2.3	25.1	2.6
Rep(−)	1.1	45.1	3.4	1.5	31.5	3.0
Rep(±)	1.7	113	4.0	2.0	62.5	2.9

Table 2 Precision comparison of Y-axis error measured by two methods

Unit: um	Laser interferometer			Eddy current sensor		
Error type	$\delta_X(y)$	$\delta_Z(y)$	$\delta_Y(y)$	$\delta_Y(x)$	$\delta_Z(x)$	$\delta_X(x)$
Position accuracy A	11.4	31.3	108.0	14.6	31.5	92.1
Rep(+)	1.7	26.4	65.3	3.5	20.6	34.2
Rep(−)	1.1	18.9	63.7	4.2	18.6	23.1
Rep(±)	1.7	27.2	67.0	3.1	19.3	31.5

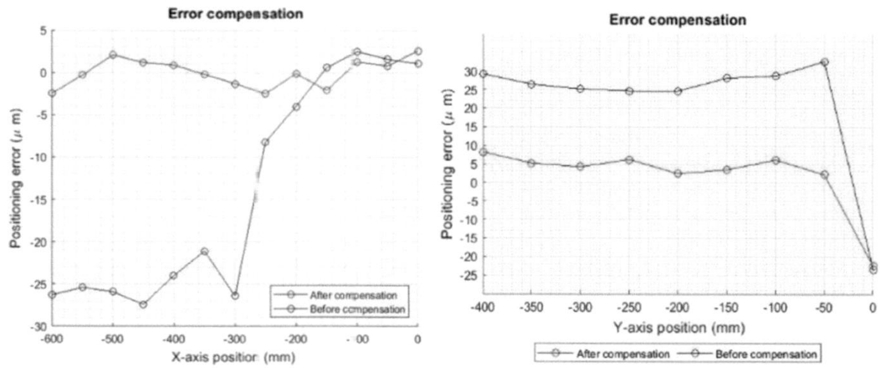

Fig. 9 Straightness error compensation results

6 Conclusion

The orthogonal tool-mounted sensor and ball array measurement artifact developed in this study, combined with the homogeneous transformation matrix, can be used to establish a volumetric error model for the entire machine tool. In the future, if the material of the ball array measurement artifact can be changed to one with a low thermal expansion coefficient, for example, Invar alloy, and manufactured with greater precision, and regular calibrations using a CMM equipment are conducted, the accuracy of the measurement results can be improved. Regarding the coils, this study has demonstrated that the eddy current coils used offer high resolution and have considerable flexibility in terms of probe size and measurement distance, with minimal environmental influence. Future improvements could include using ceramic substrates for the coils and modularizing the entire measurement system. Commercializing low-cost eddy current sensors could enhance the overall precision of machine tools.

References

1. Fan JW, Guan JL, Wang WC, Luo Q, Zhang XL, Wang LY (2002) A universal modeling method for enhancement the volumetric accuracy of CNC machine tools. J Mater Process Technol 129(1–3):624–628
2. Fan KC, Chen HM, Kuo TH (2012) Prediction of machining accuracy degradation of machine tools. Precis Eng 36(2):288-p298
3. Huang JT, Yao YC (2022) Three-axis tool embedded geometric measurement system. IN: The 19th international conference on automation technology, Taipei, Taiwan
4. Wang HB, Liu YB, Li W, Feng ZH (2014) Design of ultra-stable and high-resolution eddy-current displacement sensor system. In: 40th annual conference of the IEEE industrial electronics society, Dallas, TX, USA
5. Zhu SW, Ding GF, Qin SF, Zhuang L, Yan KY (2012) Integrated geometric error modeling, identification and compensation of CNC machine tools. Int J Mach Tools Manuf 52(1):24–29

6. Breitzke A, Hintze W (2022) Workshop-suited geometric errors identification of three-axis machine tools using on-machine measurement for long term precision assurance. Precis Eng 75:235–247
7. Xie Z, Lu Y, Chen X (2020) A multi-sensor integrated smart tool holder for cutting process monitoring. Int J Adv Manuf Technol 110(3):853–864
8. Ouyang JF, Jawahir IS (1995) Ball array calibration on a coordinate measuring machine using a gage block. Measurement 16:219–229. SSDI 0263-2241(95)00035-6
9. Renishaw (2020) XL-80. User Guide. www.renishaew.com/xl80
10. ISO Copyright Office (2014) ISO 230-2:2014 Test code for machine tools—Part 2: Determination of accuracy and repeatability of positioning of numerically controlled axes. International Standard, May (2014)

Optimization of the Production of Extruded Maca Flour Through Process Automation in an Agro-industrial Company

Jean Piero Mandujano Palacios, Mayra Nicole Castro Huamani, and Rodolfo Antonio Chávez Castillo

Abstract The Indecopi's Directorate of Distinctive Signs (DSD), highlights the quality of Peruvian maca, grown in the Andes in red, black and yellow varieties. According to the Ministry of Agrarian Development and Irrigation (MIDAGRI), with 3,201 hectares of cultivation, 18,359.92 tons are produced, valued at 11.96 million soles, which makes it an attractive option for companies. This study focuses on optimizing the production of extruded tricolor maca flour in "Super Naturals Food SAC". A Process Operations Diagram (DOP) was developed, providing a graphical representation of each operation, which facilitates the understanding of the production sequence. To improve efficiency, tools such as the Ishikawa Diagram, to analyze root causes of problems and Value Stream Mapping (VSM), comparing the current state with an ideal one, were employed. This allowed the identification of problems such as lack of monitoring in extrusion and inadequate moisture control. A machine with a temperature sensor was introduced, achieving a gelatinization of more than 94%. A cooling tunnel with humidity sensors and an automated packaging machine were also installed. In addition, conveyors were implemented for a continuous flow. The improvements reduced the loss from 8 to 0% and doubled the production capacity, reducing the manufacturing time of 500 kg of flour from 14 to 7 days, increasing annual income by S/8,789.28. The controls ensure consistent high-quality production, improving the company's competitiveness.

Keywords Tricolor extruded Maca flour · Value stream mapping · Decline · Optimization · Processes · Automation first section

J. P. M. Palacios (✉) · M. N. C. Huamani · R. A. C. Castillo
Department of Industry, Universidad Continental, Huancayo, Peru
e-mail: 76012351@continental.edu.pe

M. N. C. Huamani
e-mail: 76347685@continental.edu.pe

R. A. C. Castillo
e-mail: rchavezc@continental.edu.pe

S. Y. Liang (ed.), *Manufacturing Technologies*, Lecture Notes in Mechanical Engineering, https://doi.org/10.1007/978-3-031-95455-9_6

1 Introduction

Peru has 10 appellations of origin recognized by Indecopi's Directorate of Distinctive Signs (DSD), noted for their high quality; among them are maca, grown in Pasco and Junín [1]. Maca (Lepidium meyenii Walpers or Lepidium peruvianum Chacón) is an edible root grown in the Andes Mountains, between 2,439 and 4,419 m above the sea level, in cold climates (4–7 °C). There are three varieties of maca: the red one, with a high content of antioxidants such as carotene and niacin; the black one, noted for its speed and power, improving physical and sexual endurance, fertility and mental concentration; and yellow, which reduces stress, fatigue and acts as a natural stimulant [2]. Maca is rich in carbohydrates, providing great energy capacity. It contains 12% excellent quality proteins, with a balanced amino acid profile. It is also rich in minerals such as calcium, phosphorus, potassium, magnesium and iron. Its vitamins include thiamine (B1), essential for the nervous system, riboflavin (B2), important for vision, and vitamin C. In addition, its fiber content (cellulose and lignin) can reduce the risk of colon cancer [3, 4].

In 2023, the Ministry of Agrarian Development and Irrigation (MIDAGRI) prepared a productive profile of maca in Peru, noting that more than 3,201 hectares were cultivated and 18,359.92 tons were produced, generating a value of 11.96 million soles. Pasco and Junín are the main producers [5]. In 2021, Peru was the world's second largest exporter of maca flour and the fourth largest supplier in the US [2]. In 2023, it exported 40,234,500 kg, with an FOB value of 1,992,571.71 dollars, increasing compared to 2022. The main destinations were Ecuador, the US, Colombia and Indonesia [6]. It focuses on gelatinized maca flour, which improves solubility and digestibility [7]. The optimization of extruded maca flour production seeks to ensure quality, reduce waste, and expand the market through automatic control [8, 9].

In Ecuador, the company RistokCacao S.A. is dedicated to the export of cocoa. An analysis of the level of automation revealed that approximately 40% of the company's processes are automated, and it was also identified that the pace of production depends largely on the drying process on clotheslines. To improve this situation, the implementation of dryers on clotheslines was proposed, which would significantly reduce drying time, prevent product loss and improve product quality. When applying the traditional drying method, damage to the cocoa bean is evident, resulting in low-price sales in the national market. However, with the implementation of automation, it is hoped to eliminate damage to cocoa beans, making the product suitable for export and generating greater economic benefits [10].

A Peruvian passion fruit nectar company has a production process that consists of ten stages, of which standardization has been identified as the bottleneck. This stage is performed manually by an operator, which causes delays in subsequent operations. To solve this problem, the implementation of an automatic machine equipped with sensors and valves controlled by a PLC was proposed. Automating this process resulted in an increase in productivity of 3 bottles per hour per operator, which translates into a monthly increase of an additional 7,862 bottles [11].

In a Thai noodle production company, the cooking and packaging processes have shortcomings that affect the quality of the product. The cooking process generates excessive waste, compromising the objectives. To address this, a PLC with sensors was implemented to monitor temperature, pressure, and humidity, reducing drying time and improving quality. In the packaging, weight variations will be observed in the bags, so an automatic scale will be incorporated to ensure an even distribution. Vacuum packaging and sealing is inefficient and lacks quality control. Therefore, it is proposed to implement a horizontal automatic sealing machine that will reduce the vacuum sealing time, improving the efficiency of the production system and its responsiveness. These improvements seek to optimize processes and guarantee a higher quality product [12].

The evaluated research prioritizes improving product quality, reducing operational losses, and increasing productivity through automation. This study will simulate the implementation of automation in "Super Naturals Food SAC", focusing on extrusion, cooling and packaging processes. The extrusion temperature will be controlled with coolants, cooling will be optimized by an air injection system, extractor and humidity sensor, and packaging will be accelerated using an automated machine. These advances address challenges such as eliminating bottlenecks, reducing shrinkage to minimize economic losses, and increasing production capacity in the face of growing demand. The study seeks to create a controlled environment that optimizes process conditions through automation, thus ensuring consistency and quality in the production of extruded maca flour. These improvements are essential to respond to market demands and ensure competition.

2 Materials and Methods

This study analyzes the activities in the production of extruded tricolor maca flour in the company "Super naturals food S.A.C.". First, the activities from the transport of maca to the packaging of extruded maca flour are examined. They are considered critical processes; the extruding of the maca, the cooling of the extruded maca and the packaging of the final product.

2.1 Activities for the Production of Extruded Tricolor Maca Flour

The production of extruded tricolor maca flour at the company "Super naturals food S.A.C." includes the following stages.

Transfer of the maca to the Match area. The conditioned tricolor maca grains are transferred in 50 kg bags to the area of material in process for the match activity.

This task, carried out by 2 operators with a cart, takes approximately 20 min and allows 10 bags to be moved.

Milling process of maca. The activity begins with the feeding of the maca grains into the hopper of the milling machine, which splits them into sizes from 12 to 14 mm. Then, they are sifted and deposited in 70 kg bags. This task is carried out by 2 operators in an estimated time of 16 h.

Mixing of maca and cornstarch. The bags of split maca are transferred to a stainless steel table to be poured and then 5% of cornstarch is adhered to be mixed manually; this mixture is reserved in plastic containers. This activity is carried out by 2 operators in the estimated time of 4 h.

Extruding of the mixture. The reserved mixture is entered into the feed hopper of the extruder machine to cook the maca, which is expelled in the form of pellets. This activity is carried out by 3 operators in an estimated time of 10 h.

Cooling the mixture. The cooked maca is transferred to a stainless steel table to be laid and cooled to room temperature. This activity is carried out with the help of 2 operators and is carried out in parallel to the extrusion process.

Refining the mixture. The cold extruded maca is transferred to the mill's feed hopper, which is refined through a 5 mm to 8 mm sieve to obtain the extruded tricolor maca flour and is deposited in sacks. Two operators participate in this activity and it requires an estimated time of 16 h.

Packaging of the flour. The bags of extruded maca flour are conditioned for the weighing, packaging and sealing of the product (Doypack 250 g). This activity requires 3 operators and an estimated time of 4 days.

Packaging the individual boxes. The 250 g Doypacks are packaged in individual boxes of folcote together with a measuring spoon. This activity requires 3 operators and an approximate time of 3 days.

Packaging of the final product. The packaged product is packed in boxes with a capacity of 37 units and then packaged and labeled. This activity is carried out by 2 operators and is carried out in parallel to the packaging process of the final product.

Figure 1 shows the graphic representation of the processes involved in "Super naturals food".

Figure 2 shows the processes of "Super naturals food S.A.C." for the production of extruded tricolor maca flour.

2.2 Analysis of Critical Processes in the Production Process

Extruded tricolor maca mixed with cornstarch. The company "Super naturals food S.A.C." cooks the split maca taking into account the technical specifications of foods that are part of the provision of the food service of the national school feeding

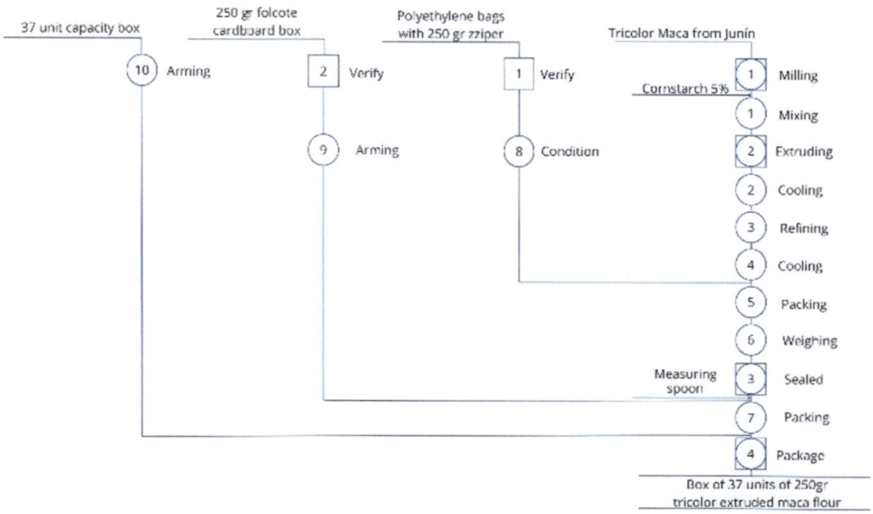

Fig. 1 DOP for the production of extruded tricolor Maca flour

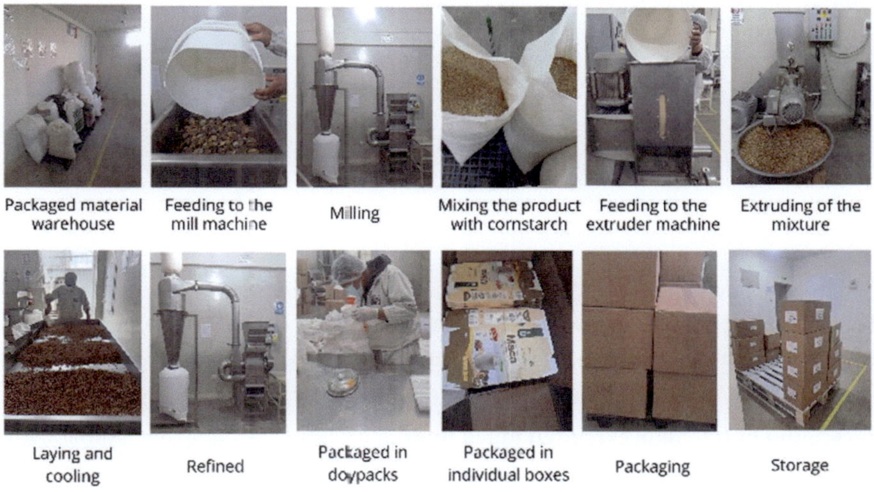

Fig. 2 Production processes of extruded tri-color Maca flour

program Qali Warma, which indicates that the gelatinization rate must be greater than 94% [13]. To reach the established index, the cooking temperature must be between 90 and 110 °C.

Table 1 "Super Naturals Food SAC" cooking temperature per batch

Day	Average temperature (°C)	Gelatinization index (%)
1	95–100	96–97
2	90–95	94–95
3	105–110	100
4	85–90	92–93
5	85–90	92–93

Table 1 shows that 40% of the data obtained are below the acceptable gelatinization range; which directly affects the quality of the product, being a control problem for the company.

Cooling of the extruded tricolor maca. This process is carried out through cooling at room temperature, which is often variable with respect to the cooling time because a humidity percentage of less than or equal to 5% must be met. Table 2 shows the analysis carried out on the cooling of 500 kg of extruded tricolor maca during 5 days of production, highlighting the variable of the percentage of moisture.

Table 2 "Super Naturals Food SAC" cooking temperature per batch

Day	Average humidity (%)
1	4–5
2	6–7
3	7–8
4	4–5
5	6–7

Table 2 shows that 60% of the total average humidity measurements are above the established parameters, which is detrimental to the quality of the product. It is also known that to comply with these parameters 7–8 h are required; which results in a delay in production.

Packaging of tri-color extruded maca flour. Currently, the company "Super naturals food S.A.C." carries out this process manually, so it generates an average waste of 4% per day, this was calculated after analyzing the data obtained during 10 days of evaluation.

According to Fig. 3, it can be seen that the process that generates the greatest loss is packaging, which presents 4% of waste equivalent to 20 kg, being 50% of the total waste generated. This wasted product causes associated costs that are presented in Table 3.

It is observed that the company "Super naturals food S.A.C." generates 20 kg of waste in the packaging process per day of production, the loss generated is equivalent to the loss of 80 units of extruded tricolor maca flour in the presentation of 250 g, which are sold at the price of S/. 20 soles; therefore, the company has sales losses equivalent to S/. 1600 soles per day of production.

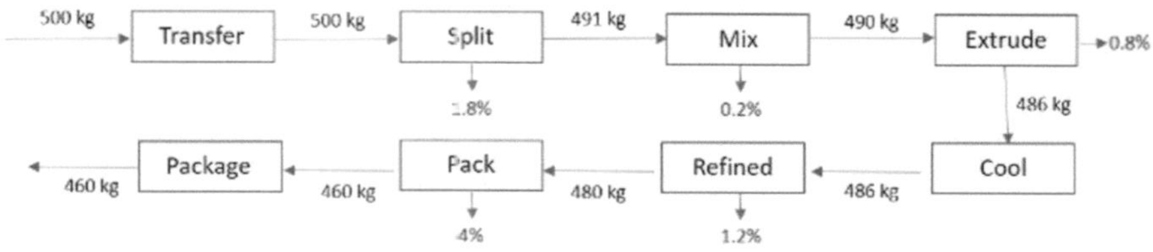

Fig. 3 Losses generated in each manufacturing process

Table 3 Loss of sales due to the decline of "SAC Supernatural Foods"

Kilograms of loss per day (kg)	Units lost due to lost working hours (units)	Unit sale price of the final product (S/.)	Lost sales (S/.)
20	80	20	1600

2.3 *Evaluation of Indicators Using the Ishikawa Diagram*

The Ishikawa diagram, or "fishbone," was developed by Dr. Kaoru Ishikawa in the 1960s. This tool visualizes the possible causes of a problem, helping to identify its root causes. With the problem in the head and the causes spreading like branches from the body, it makes it easier to explore sub-causes and identify the most likely factors originating the problem [14].

Figure 4 details the causes and problem identified, this diagram was made for the analysis of the causes of the deficient production process of tricolor maca extruded flour, using the classification of the 6 M, identifies key areas for improvement, which allows identifying specific actions to optimize processes, reduce waste and improve efficiency (Fig. 4).

2.4 *Value Stream Mapping (VSM) Analysis of "Supernatural Foods SAC"*

Value Stream Mapping (VSM) is a key tool in Lean Manufacturing that allows the visual representation and optimization of processes. This technique is effective in shortening lead times and reducing costs by identifying and eliminating activities that do not add value. A value stream map of the current state is generated, visualizing the non-value activities. Then, a map of the future state is made, eliminating activities that generate waste [15, 16]. Figure 5 presents the analysis carried out using the VSM, which helps to identify areas for improvement to optimize the production process of extruded tricolor maca flour.

To elaborate the Value Stream Mapping (VSM) diagram, data were collected through direct observation and interviews with workers in the production process. One of the authors, linked to the company as an employee, had privileged access to

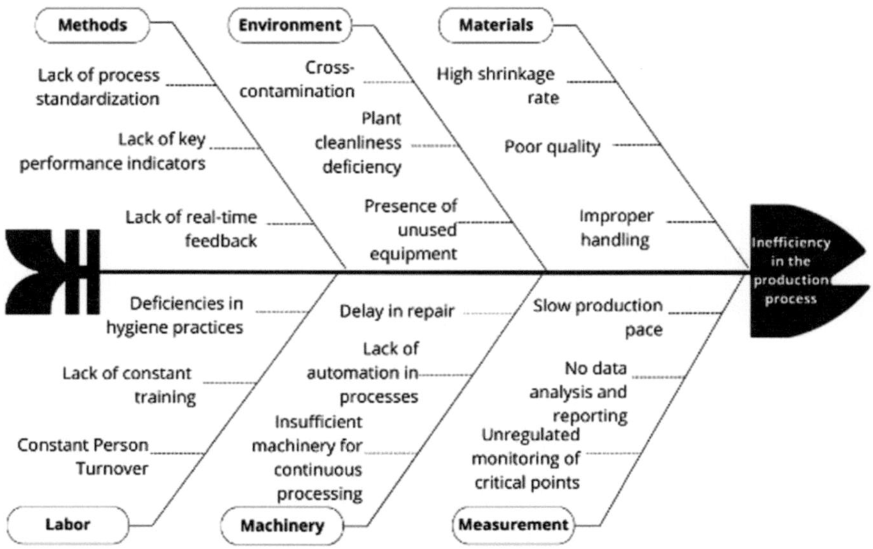

Fig. 4 Company Ishikawa diagram "Super Natural Foods SAC"

internal operations, facilitating data collection and allowing key information to be obtained for analysis. This level of involvement led to a deeper understanding of the dynamics of the process and the identification of inefficiencies and opportunities for improvement in the value chain.

3 Results

The results obtained are based on the requirements analyzed through Value Stream Mapping (VSM), which proposes optimizations through the Lean Manufacturing methodology. Additionally, simulations of the necessary environments for the company "Super Naturals Food S.A.C." are carried out.

3.1 Manufacturing Improvement Plan—VSM

Lean Manufacturing is an integrated socio-technological system designed for process improvement, with the fundamental purpose of eliminating waste or activities that do not add value to the customer. The elimination of this waste leads to an increase in quality, as well as a significant reduction in production times and costs in a relatively short period.

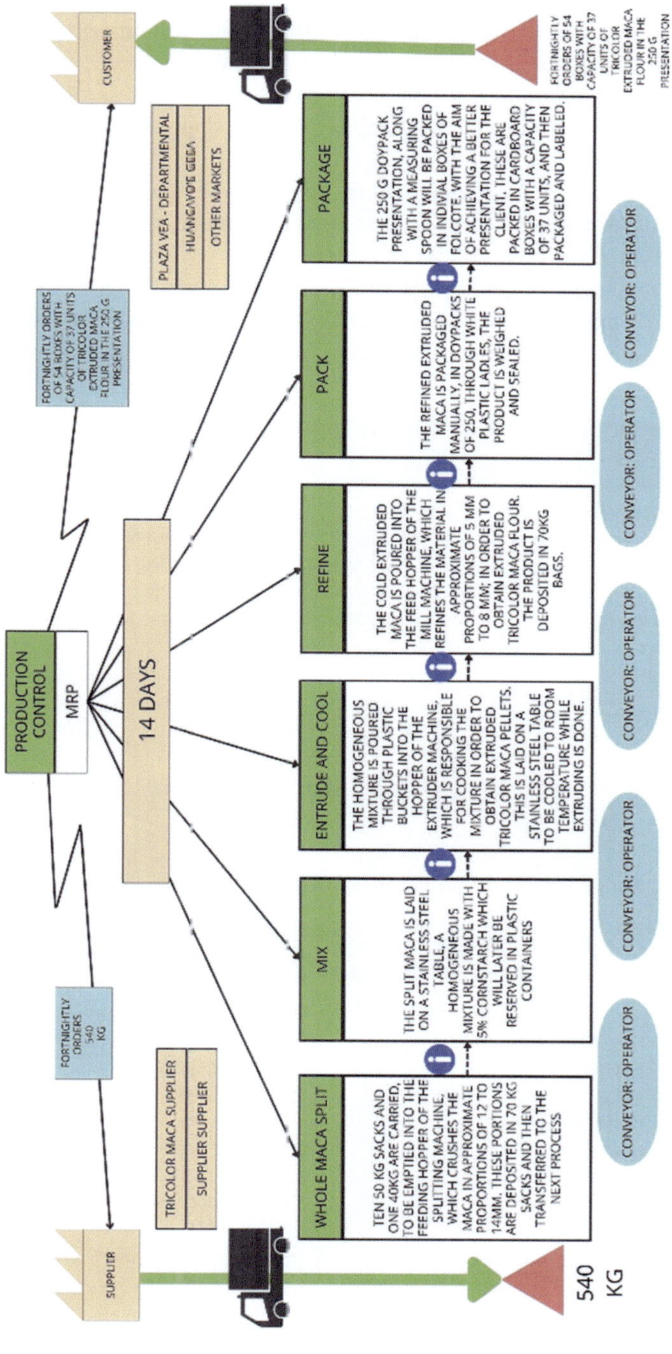

Fig. 5 Current diagram of the VSM of the company "Super Naturals Food S.A.C."

This entails a detailed assessment of the value stream to detect opportunities for improvement, carried out through value stream mapping (VSM). The purpose is to reduce the loss of time and resources during the production process as much as possible in order to increase the value of the final product [17].

The diagnosis, operation and follow-up to improve the process of production of extruded tricolor maca flour in the company "Super Naturals Food S.A.C." are presented in Fig. 6.

To this end, organizational requirements, personnel related barriers and techniques for the implementation of improvements in the processes of the production of extruded tricolor maca flour will be identified.

Table 4 present a detailed description of the processes in the improvement phase in the company "Super Naturals Food S.A.C.", after a thorough evaluation of the value stream mapping (VSM) and a critical analysis of the Ishikawa Diagram.

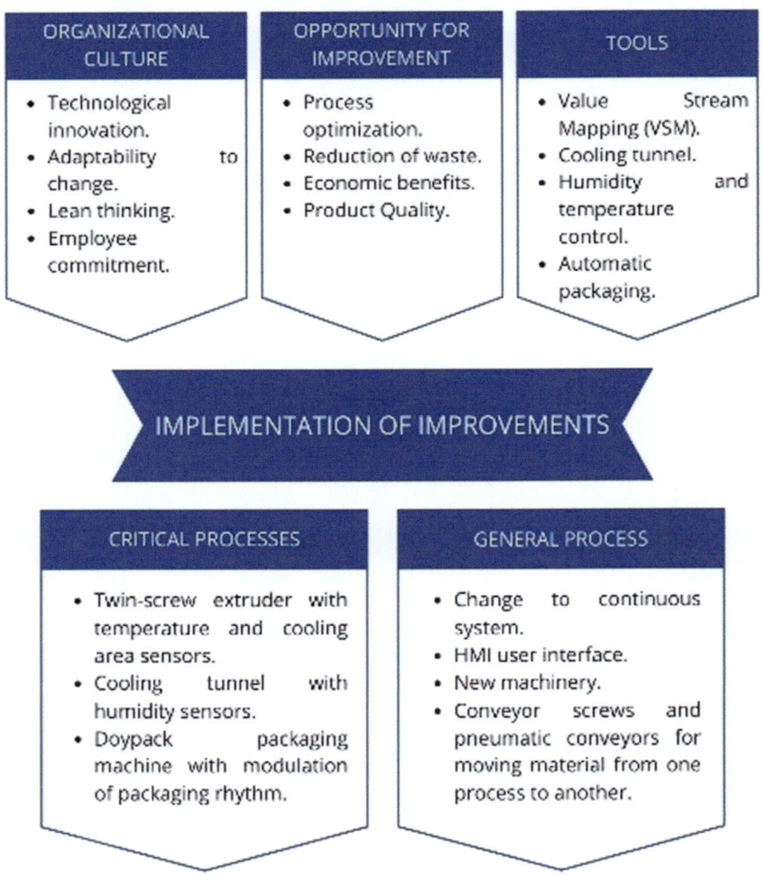

Fig. 6 Views of the operation of the proposed machine

Table 4 "Super Naturals Food SAC" Cooking temperature per batch

Critical process	Description
Extrusion	The material is transferred by means of a conveyor screw to the hopper of the double-screw extruder, which has a temperature sensor controlled by a PLC, ensuring a gelatinization of more than 94%. It includes automatic coolants and a cruise control, with data visible on an HMI interface
Cooled	The material reaches the cooler via a pneumatic conveyor. This cooler, with a circular drum that rotates at 360°, cools the granulated maca by cold air in opposite flow and extracts hot air. A sensor controlled by a PLC monitors the humidity of the product, keeping it below 5%, with data displayed on an HMI interface
Packing	The pneumatic conveyor pours the material into the hopper of the 250 g doypack packer. This automated machine conditions the container with pacifiers for the filling of extruded tricolor maca flour, controlling the feed according to the required weight. Each container is sealed and automatically transferred by bands to a tank, optimizing a process that currently lasts 4 days

These improvements are aimed at optimizing production efficiency by reducing non production times and operational losses, in order to improve product quality and ensure a continuous and profitable process in the manufacture of extruded tricolor maca flour.

These improvements aim to optimize production efficiency by reducing downtime and operational losses, thereby improving product quality and ensuring a continuous and cost-effective process. Figure 7 shows the future VSM with the improvements implemented.

3.2 *Graphical Simulation of What Is Required by the VSM*

The HMI user interface displays the design and simulation of all the necessary processes, developing three different screens that allow comprehensive monitoring of the system. Figure 8 illustrates a panel where all sensors and actuators involved in the extrusion process can be monitored, as well as displaying the warning or error messages of the proposed system This interface also includes representations of the cooling process, allowing to visualize the percentage of moisture of the product and automatically issuing warnings if the parameterized limits are exceeded, as well as the packaging system, where the stages of the process, the number of packaged units and the production rate in units per minute are evidenced, giving the user the ability to select and control the desired packaging.

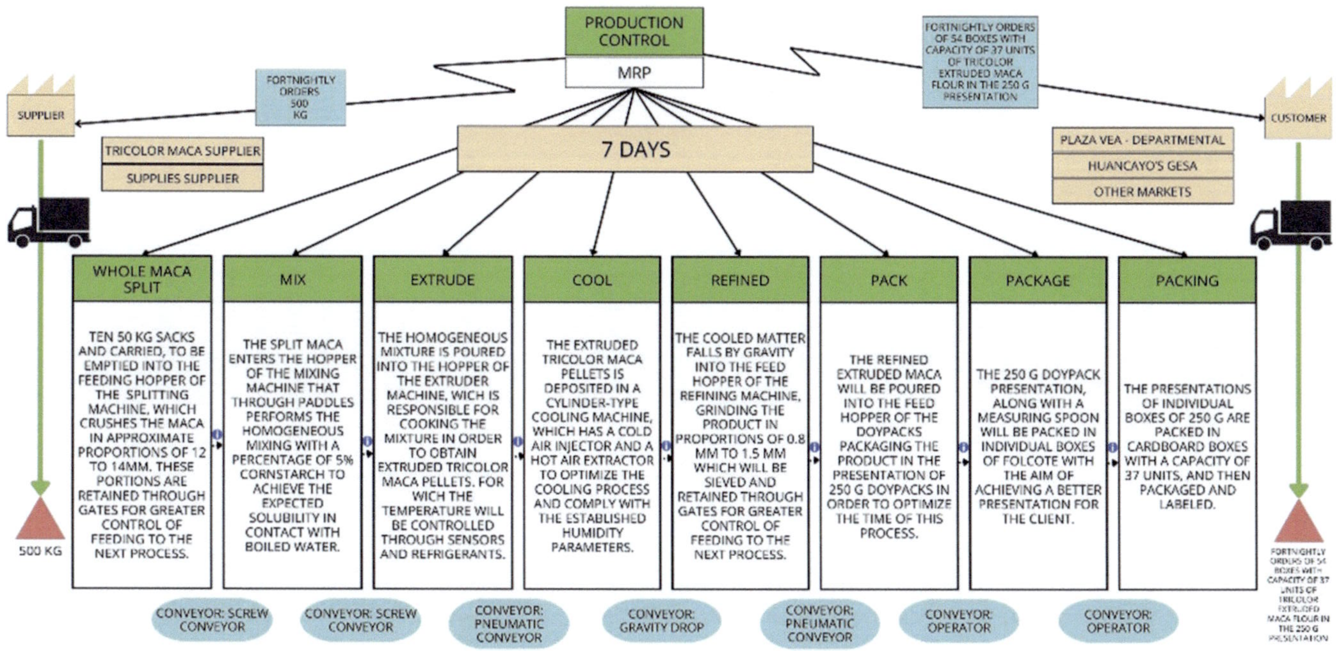

Fig. 7 Future VSM diagram of the company “Super Naturals Food SAC”

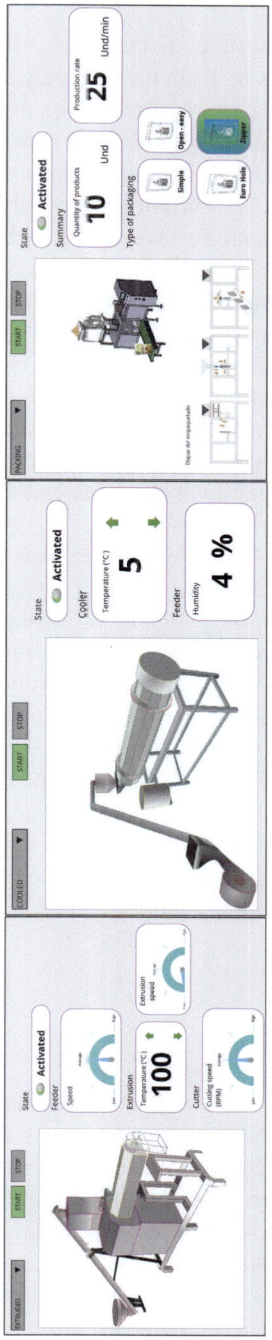

Fig. 8 Visualization of the extrusion process

3.3 Cost Analysis

After the implementation of the system, a significant reduction in the operational shrinkage rate is expected, decreasing from the current 8% to 0%. This is due to the fact that the new system no longer requires direct labor, avoiding operational errors, and the redesign of the process that includes new machinery, which allows a continuous flow of material between processes and precise control of quality parameters such as temperature and humidity. The current operational losses generate economic losses for the company, since if this product were marketed, greater income would be obtained. In addition, the proposed system would reduce labor costs. A 50% reduction in production time is anticipated, similar to the 40% mentioned by Cruz [18] in downtime and shrinkage. This improvement is achieved with the implementation of a continuous system with pneumatic and screw conveyors, avoiding stops and overhandling. Currently, the automatic packaging machine is the slowest process, while the cooling tunnel speeds up production. Making 2000 units of extruded tricolor maca flour in a 250 g presentation currently requires 14 days; With the proposal, it is expected to reduce this time to 7 days, which will decrease labor costs and increase profitability by increasing production capacity.

Table 5 provides an exhaustive breakdown of the costs associated with operating losses, which could have been marketed in units if the proposed system had been implemented. The costs of inputs, the acquisition of tricolor maca and labor were included according to the data provided by "Super Naturals Food". Likewise, the price of cornhouse is to be S/6.38 per kilogram according to MIDRAGRI's Agro Chatea platform [19]. By requiring only one operator to manage the proposed system, a minimum wage of S/1025 was estimated, based on the processing of 500 kg of extruded tricolor maca flour. This would generate a net income of S/2.05 per kilogram processed on a monthly basis. In terms of kilograms lost, this would be equivalent to a total annual income of S/8789.28 nuevos soles. It is important to mention that additional expenses such as water, electricity, and space used were excluded in this analysis, since the focus was on primary inputs.

4 Discussions

After reviewing the bibliography, none of the antecedents found focus on the same product; therefore, it was considered important to compare this work with a study on the optimization of food production processes, specifically the one carried out by L. Baque, M. Cadena, D. Salvatierra and K. Quiñonez [10]. These authors propose automating 50% of processes, seeking to increase existing automation by 10% to 40%. This study, on the other hand, proposes a total redesign of the production area, increasing automation to 80% to achieve a continuous flow of material. Both studies agree that no personnel are needed for each process, as automation allows for efficient control. In addition, it seeks to improve the quality of the product, generating

Table 5 Cost analysis after implementing the streamlined continuous flow system

Article	Weight (kg/und)	Cost per corn house acquisition	Cost of acquiring Doypack	Cost of acquiring folcote box	Unit cost of product sales (kg)	Maca acquisition cost (kg)	Income from the sale of maca	Net income	Number of times per year	Annual net income
Maca	40/160	S/ 12.76	S/ 76.8	S/400	S/20	S/ 8.50	S/ 800	S/ 732	12	S/ 8,789.28

economic benefits through a drying system that optimizes times and reduces damage. This study implements an extrusion machine with temperature sensors, a cooling machine with moisture sensors and a packaging machine, ensuring that the product meets the standards required for consumption.

In the study by A. Siapo [11] in a passion fruit nectar plant, an integral automation is proposed that replaces the containers in the standardization process with a machine with sensors and valves controlled by a PLC, optimizing the precision and efficiency of the process and improving quality control. Humidity and temperature sensors managed by the PLC will be installed for continuous real-time monitoring. A Human Machine Interface (HMI) will also be implemented to facilitate interaction between the operator and the machine, improving operability and responsiveness to variations. Both studios seek to increase production through automation. The Peruvian studio increases its production by 7,862 bottles, generating a profit of S/ 141,516. In contrast, this study increases production by 160 units of extruded maca flour, generating a profit of S/ 8,789.28.

5 Conclusions

In conclusion, when analyzing the production processes of extruded tricolor maca flour in the company "Super Naturals Food S.A.C.", the importance of tools such as the Ishikawa diagram and the Value Stream Mapping is highlighted. These tools are critical for identifying the underlying causes that can negatively impact the efficiency of activities. The implementation of these diagnoses allows for the design and increase of actions aimed at improving the current state of the process.

The implementation of a continuous flow system in controlled environments represents a significant advance in the production of tri-color extruded maca flour. This approach reduces the shrinkage rate from the current 8% to almost 0%, increasing production capacity and improving demand response. These benefits translate into substantial economic improvements. A cost analysis shows that reducing operational waste has a direct impact on profitability. By reducing losses, the use of resources is optimized, which allows the volume of production available for sale to be increased. It is estimated that the annual net income from the commercialization of extruded flour could reach S/ 8,789.28, evidencing the efficiency of automation in reducing losses and optimizing processes, contributing to the economic sustainability of the company. Automation also improves the quality of the final product and strengthens competitiveness in a demanding market.

It is important to highlight that the automated system proposed for the production of extruded tricolor maca flour at "Super Naturals Food" incorporates a controlled environment that simulates the optimal conditions for processing, including the precise regulation of temperature, humidity and other critical parameters. Although the system demonstrates significant improvements in efficiency and reduction of waste, it is important to recognize the potential limitations related to its integration into existing facilities. These considerations are essential to evaluate the return on investment and viability of the system in the context of actual plant operations.

Finally, several constraints and challenges for the future must be considered. Although the proposed system shows great potential, it faces limitations in its performance in a real production environment. Integrating this technology into existing infrastructure can present technical challenges. In addition, the current results are based on a simulated virtual environment, which does not always reflect the conditions of the real environment. It is recommended that future studies include the area of logistics, to obtain a more complete view of the impact of the system on production efficiency and inventory management, allowing to identify areas for improvement and guarantee benefits throughout the supply chain.

References

1. National Institute for the Defense of Competition and the Protection of Intellectual Property (INDECOPI). https://www.gob.pe/institucion/indecopi/noticias/809340-orgullo-peruano-conoce-nuestros-10-productos-con-denominacion-de-origen-que-son-motivo-de-celebracion-este-28-de-julio. Last accessed on 19 July 2024
2. Market Intelligence Department: Market Product Profile: Maca powder in the United States. https://boletines.exportemos.pe/recursos/boletin/perfil%20producto%20mercado%20-%20maca%20en%20polvo%20en%20estados%20unidos.pdf. Last accessed 25 July 2024
3. Aliaga R, Espinoza E, Rodríguez G, Villagómez V, Jananpa M, Bazán R, Blas R, Llanos N (2011) The maca value chain in the Meseta del Bombón: analysis and strategic guidelines for its development. Lima
4. Ministry of Agriculture and Irrigation (MINAGRI). https://repositorio.midagri.gob.pe/handle/20.500.13036/302. Last accessed on 22 July 2024
5. Ministry of Agrarian Development and Irrigation (MIDAGRI). https://app.powerbi.com/view?r=eyJrIjoiN2U3NTdkYTktOTk5Ny00NjQ5LTg0ZjEtMmIzYzlmZWIwMDhlIiwidCI6IjdmMDg0NjI3LTdmNDAtNDg3OS04OTE3LTk0Yjg2ZmQzNWYzZiJ9. Last accessed on 25 July 2024
6. National Superintendence of Customs and Tax Administration (SUNAT). http://www.aduanet.gob.pe/cl-ad-itestadispartida/resumenPPaisS01Alias. Last accessed on 30 July 2024
7. Camavilca U, Camavilca U (2012) Feasibility study for the installation of a processing plant for gelatinized maca flour (Lepidium peruvianum Chacón) for export to the United States
8. Tecún H (2020) Internal control of manuals of procedures of good manufacturing practices for safety in the food industry in Mixco, Guatemala
9. García R, Mera P (2022) Automation of a micro ingredient weighing system for the production of balanced pet food. ESPOL. FIEC, Guayaquil
10. Baque L, Cadena M, Moserratte D, Quiñonez K (2020) Analysis of the levels of automation of cocoa production processes. J Bus Stud Entrep 4(2):8–23
11. Siapo A (2022) Automation proposal to improve the productivity of a passion fruit nectar producing company
12. Siwasit P, Parida J, Pinit N (2024) Improving Lean-Kaizen practices through IoT and automation: a comprehensive analysis with simulation models in the Thai food industry. Res Eng Appl Sci 51(3):286–299
13. Ministry of Development and Social Inclusion (MIDIS). https://info.qaliwarma.gob.pe/datpub/uop/catalogo/2023/31-HARINAS-EXTRUIDASFF.pdf?v=2.0. Last accessed 06 August 2024
14. García E (2024) Ishikawa diagram: a tool to identify and solve problems in the agri-food industry

15. Sheth P, Deshpande V, Hiren K (2014) Value stream mapping: a case study of the automotive industry. Int J Res Eng Technol 3(1):310–314
16. Joshi R, Naik G (2012) Process improvement using value stream mapping: a case study in the small-scale industry. Int J Res Eng Technol 1(5):1–10
17. Tejeda A (2011) Improvements of lean manufacturing in production systems. Sci Soc 36(2):276–310
18. Cruz S (2023) How to reduce downtime in industries with continuous production? https://thefoodtech.com/columnistas/como-reducir-el-tiempo-muerto-en-industrias-con-produccion-continua/. Last accessed on 05 October 2024
19. Ministry of Agrarian Development and Irrigation (MIDAGRI). https://siea.midagri.gob.pe/portal/agrochatea. Last accessed on 15 August 2024

Investigation on Damping Properties of an Elastic Ring Squeeze Film Damper Considering Contact Deformation

Yangpeng Li, Shuyun Jiang, Yuliang Lu, and Yunan Zhou

Abstract The damping properties of elastic ring squeeze film damper (ERSFD) is obscure, particularly the causes and effects of elastic ring deflection. On account of the theory of fluid film lubrication, the central difference method is utilized to calculate the ERSFD dynamic properties considering contact deformation, the effects of convexities and ER deflection on the oil film pressure are analyzed, and the effect of eccentricity on the mechanical properties of ERSFD is discussed. The results show that the contact deflection reduces the maximum value of oil film pressure and makes the oil film peak sparse, and the unique factor in the generation of external oil film pressure is contact deformation.

Keywords Elastic ring deflection · Damper · Damping properties · Contact deformation · Squeeze oil film

1 Introduction

The research and application of squeeze film damper (SFD) vibration reduction technology is regarded as a milestone in the history of aero-engine development [1]. However, the disadvantage of the non-linearity properties of the oil film force at high eccentricity severely limits the application of SFD [2]. To improve the drawbacks of SFD, the elastic ring squeeze film damper (ERSFD) had been invented and successfully applied to active aviation engines [3]. Figure 1 shows the structure of ERSFD, which mainly includes an elastic ring (ER), housing and journal. In order to ensure the flow of the internal and external oil film, there are some small orifices on the ER.

Y. Li · S. Jiang (✉)
School of Mechanical Engineering, Southeast University, Nanjing, China
e-mail: jiangshy@seu.edu.cn

Y. Lu · Y. Zhou
Aviation Key Laboratory of Science and Technology, AVIC Nanjing Engineering Institute of Aircraft System, Nanjing, China
e-mail: luyuliang@neias.cn

S. Y. Liang (ed.), *Manufacturing Technologies*, Lecture Notes in Mechanical Engineering, https://doi.org/10.1007/978-3-031-95455-9_7

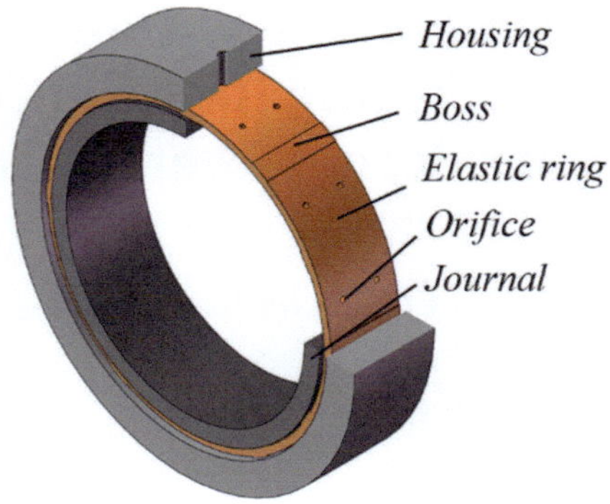

Fig. 1 Schematic diagram of ERSFD structure

Elastic ring deformation is extremely important for the mechanical properties of ERSFD. And the solution approaches mainly include approximate approach, semi-analytical approach, analytical approach and finite element approach, among which the finite element approach is accurate and widely used [2–5]. The dynamic analysis of the rotor-ERSFD system also visually demonstrates the damping reduction effect of ERSFD [6]. However, it is not easy to clearly and intuitively analyze the vibration reduction mechanism of ERSFD through the rotor system, and there are relatively few published literature. CFD software can also be utilized to investigate the dynamic properties of ERSFD [7, 8], but the calculation is time consuming.

The fluid pressure of ERSFD is not only affected by the ER convexities, but also influenced by the contact effect of journal. This paper investigates the vibration reduction mechanism of ERSFD considering the ER contact deformation. Based on the theory of fluid film lubrication, the internal and external oil film stresses are determined by utilizing the central difference approach. The impacts of journal contact and convexities on the oil film stress and force of ERSFD are investigated. The effect of ER deflection is studied, and the influences of journal eccentricity on the damping properties of ERSFD are investigated.

2 Theoretical Model of ERSFD

Figure 2 shows the ERSFD coordinate system and deformation diagram, here y is the vertical direction of the Cartesian coordinate system. In the figure, φ is the position angle in the rotation coordinate system, ω represents the rotational frequency, e is the eccentric distance, c_I and c_O are the internal and external oil film clearances respectively, α_i is the angle at the centerline of each internal convexity. In addition, the dashed tie indicates the deformation of this ER.

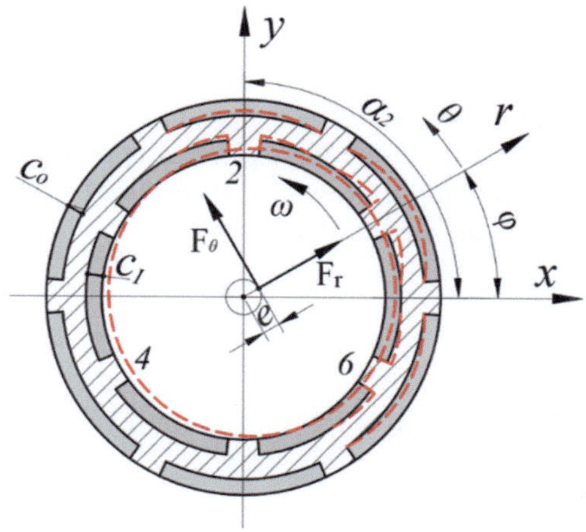

Fig. 2 Coordinate system and deformation diagram of ERSFD

2.1 *Mathematical Model of Oil Film*

The following assumptions are given for solving for the ERSFD oil film force: neglecting fluid inertia force and volume forces; neglecting the effect of the transition fillet at the ER convexity on the film thickness; ignoring the effect of small orifices of ERSFD. Based on the above assumptions, the Reynolds equations for the internal and external oil films of ERSFD can be expressed as [3, 8, 9]:

$$\begin{cases} \dfrac{\partial}{R^2\partial\theta}\left(h_I^3\dfrac{\partial p_I}{\partial\theta}\right)+\dfrac{\partial}{\partial z}\left(h_I^3\dfrac{\partial p_I}{\partial z}\right)=-12\mu_l\omega\left(\dfrac{\partial h_I}{\partial\theta}-\dfrac{\partial r_s}{\partial\theta}\right)+12\mu_l\dfrac{\partial h_I}{\partial t} \\ \dfrac{\partial}{R^2\partial\theta}\left(h_O^3\dfrac{\partial p_O}{\partial\theta}\right)+\dfrac{\partial}{\partial z}\left(h_G^3\dfrac{\partial p_O}{\partial z}\right)=-12\mu_l\dfrac{\partial r_s}{\partial t} \end{cases} \tag{1}$$

where, R is the damper radius, p_I and p_O are the internal and external oil film pressures, respectively, μ_l is the lubricant dynamic viscosity, and $r_s(\theta, t)$ is the ER deflection. The inner oil film thickness h_I and external oil film thickness h_O can be expressed as:

$$\begin{cases} h_I = c_I - e\cos\theta + r_s(\theta, t) \\ h_O = c_O - r_s(\theta, t) \end{cases} \tag{2}$$

The derivative of the thickness over time is:

$$\begin{cases} \dfrac{\partial h_I}{\partial t} = -\dot{e}\cos\theta + \dfrac{\partial r_s(\theta, t)}{\partial t} \\ \dfrac{\partial h_O}{\partial t} = -\dfrac{\partial r_s(\theta, t)}{\partial t} \end{cases} \tag{3}$$

2.2 ER Model

In this paper, the ER is regarded as thin shell elements, and the deflection of the ER is solved using the finite element approach [3]. According to the actual situation, the external protrusion is considered as a rigid contact. In the Cartesian coordinate system, the angle α_i is:

$$\alpha_i = 2\pi(i-1)/N + \pi/N \tag{4}$$

The contact effect of the journal can cause deflection of the ER, and the displacement δ_i of the internal convexity is:

$$\delta_i = \begin{cases} e\cos(\alpha_i - \varphi), & \delta > 0 \\ 0 & \text{else} \end{cases} \tag{5}$$

After calculating the deflection of the ER $r_s(\theta, t)$, the deflection speed of the ER can be expressed as:

$$\frac{\partial r_s}{\partial t} = \frac{r_s(\theta, t) - r_s(\theta, t - \Delta t)}{\Delta t} \tag{6}$$

2.3 Numerical Solution

There are analytical and numerical methods for solving the oil film stress of ERSFD. As the rise of L/D ratio, the calculation error of the analytical approach based on short bearing theory will gradually increase. Therefore, the central difference approach is investigated to calculate the oil film stress of the ERSFD, and the difference formula is expressed as:

$$A_{i,j}P_{i+1,j} + B_{i,j}P_{i-1,j} + C_{i,j}P_{i,j+1} + D_{i,j}P_{i,j-1} - E_{i,j}P_{i,j} = F_{i,j} \tag{7}$$

A difference coefficient of traditional SFD can be obtained by ignoring the deflection of the ER and the height of the convexity. For the above equation, considering the π oil film boundary condition, Matlab programming is used to solve the stress of the internal and external oil films. Radial force F_r and circumferential force F_θ of the oil film can be expressed as:

$$\begin{bmatrix} F_r \\ F_\theta \end{bmatrix} = \int_{-\frac{L}{2}}^{\frac{L}{2}} \int_{\theta_1}^{\theta_2} p(\theta, z) \begin{Bmatrix} \cos\theta \\ \sin\theta \end{Bmatrix} R\, d\theta\, dz \tag{8}$$

In this paper, a comparative analysis was conducted on the dynamic properties of the ERSFD, which ignores the contact deflection of the internal convexities and only considered the separation effect of the protrusions.

3 Results and Discussion

Table 1 provides the specific parameters of the ERSFD applied in the calculation of this paper.

3.1 *Deformation of the ER*

Figure 3 shows the deformation diagram of the ER considering contact effect at $n = 6\,\text{krpm}$, $\varepsilon = 0.2$, where the journal is located at the centerline of the inner protrusion and location angle $\varphi = 30°$. As shown in the figure, the contact effect of the journal causes significant deflection of the ER, which will profoundly affect the dynamic and static properties of the ERSFD. Secondly, when the journal position changes, the deflection of ER follows, indicating that the deflection of ER is a dynamic process.

3.2 *Inner Oil Film Pressure Distribution*

Figure 4 shows the calculated internal oil film stress distribution of ERSFD at rotational speed $n = 6\,\text{krpm}$, eccentricity ratio $\varepsilon = 0.2$, position angle $\varphi = 30°$. And the blank area in the figure indicates the position of the internal convexities. In general, the continuous oil film stress is separated into multiple oil film peaks by ER convexities. Secondly, the oil film stress considering the convexity effect and contact deformation effect is small and more sparse between the peaks. This is mainly due to

Table 1 Styles structural parameters of ERSFD

Parameters	Value
Damper length L (mm)	12
Fluid viscosity μ_l (Pa s)	0.018
Inner clearance c_I (μm)	200
Outer clearance c_O (μm)	200
Internal radius r_I (mm)	23.65
External radius r_O (mm)	25

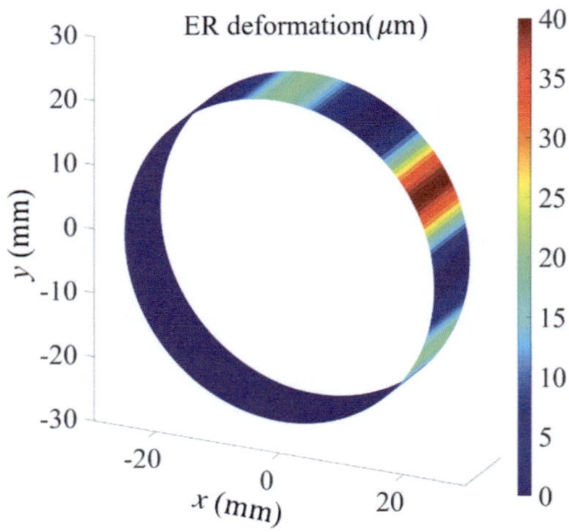

Fig. 3 The ER deformation diagram considering contact effect

the ER ring deflection from the journal contact leads to an increase in internal fluid thickness, a decrease in oil film stress, and more thinning between pressure peaks.

To better comprehend the role of the bosses and ER deflection on the inter oil film stress, Fig. 5 gives a polar schematic of internal oil film stress at the journal centreline. From the picture, the separating effect of the convexities reduces the oil

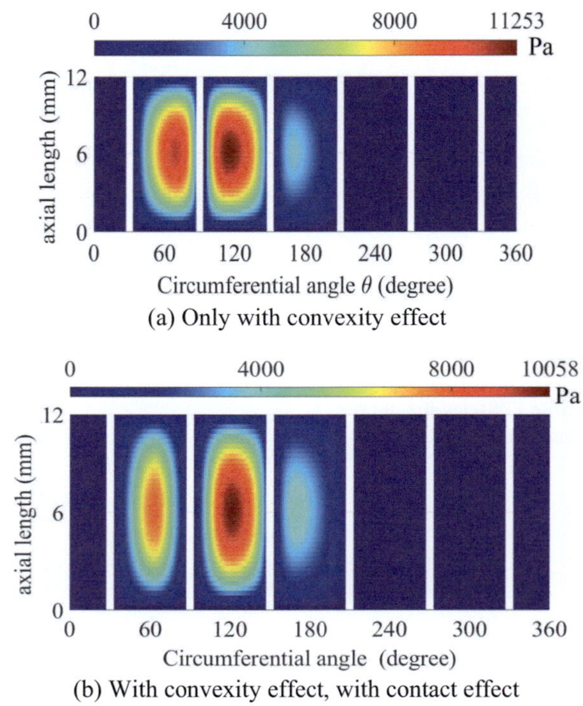

Fig. 4 Inner oil film pressure distribution

Fig. 5 Polar schematic of internal oil film pressure

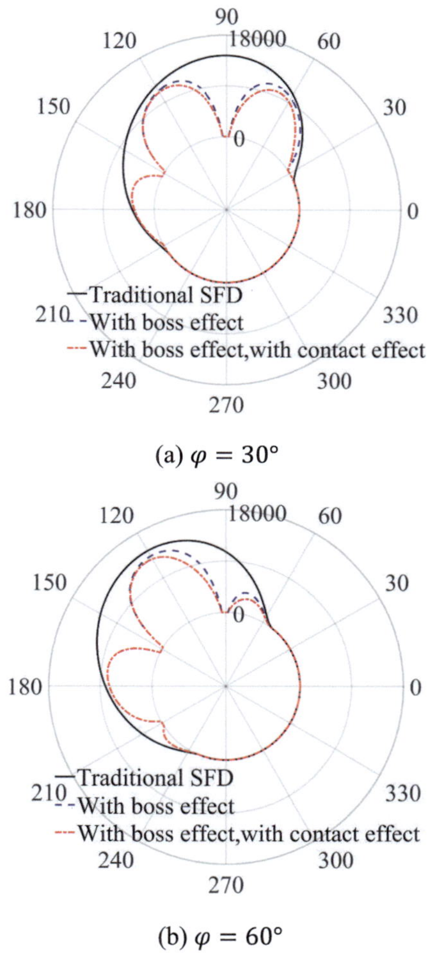

film stress of the ERSFD, after that, the contact deformation of the ER continues to reduce the internal oil film stress by increasing the film thickness. Secondly, only half of the ER deflection affects the internal oil film pressure because of the π oil film boundary condition.

3.3 Outer Oil Film Pressure

Figure 6 shows the external fluid pressure distribution of ERSFD considering the contact effect and boss effect, and there is no external oil film stress when only with the boss effect. This is because the external oil film stress is only related to the ER deflection speed $(\partial r_s/\partial t)$, which is generated only by the journal contact.

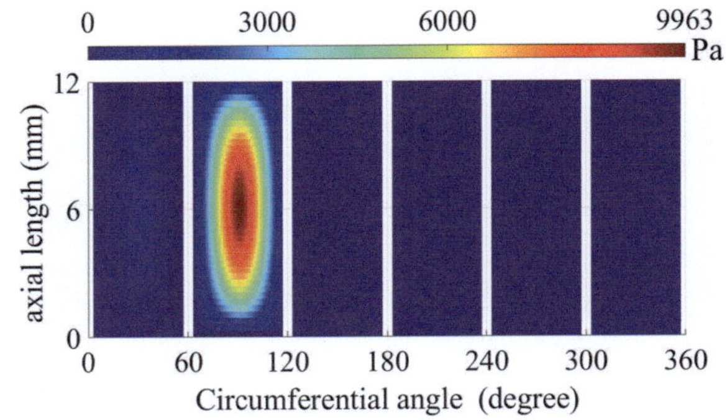

Fig. 6 External pressure distribution with contact effect and boss effect ($n = 6\,\text{krpm}$, $\varepsilon = 0.2$, $\varphi = 30°$)

Comparing Figs. 4 and 6, it was found that the external fluid stress and the inner fluid pressure are in the same order of magnitude, but the external stress does not have more stress peaks. The main reason is that the contact deformation is the only factor that generates the external oil film stress, and only when the deflection of the ER gradually increases, the external oil film pressure can be generated.

3.4 Eccentricity Ratio

Figure 7 pictures the variation curve of fluid force with eccentricity ratio at $n = 6$ krpm. In general, the fluid force of both ERSFD and SFD increases gradually with the increase of eccentricity ratio, and the amplitude of SFD is always larger than that of ERSFD.

Secondly, as the eccentricity ratio rises, the difference in the fluid force of ERSFD considering contact effect and convexity effect gradually increases compared to only with the convexity effect. This is due to the large ER deflection considering the contact effect at large eccentricity, which leads to an increase in the thickness of the internal oil film, a slow increase in the fluid force of ERSFD, and a weakening of the effect of eccentricity.

Fig. 7 Internaloil film force versus eccentricity ratio

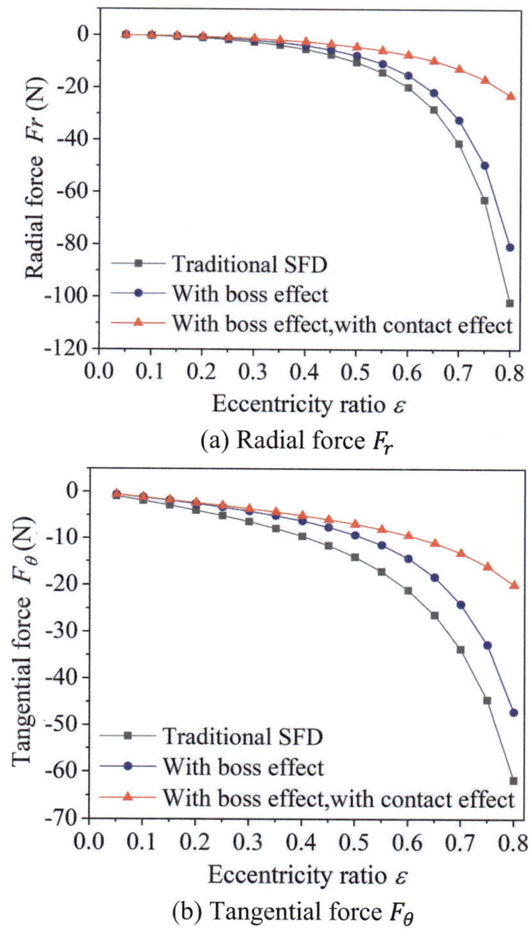

4 Conclusion

Based on the theory of fluid film lubrication, this paper establishes a dynamic model of ERSFD considering journal contact and convexities effect. The fluid pressure of ERSFD is solved by the central difference approach, the dynamic characteristics of ERSFD considering journal contact and only convexity separation action are compared and analyzed, and the damping mechanism of ERSFD is investigated. Based on the results and discussion, the following main summaries are drawn:

- The continuous inner fluid pressure is separated into multiple fluid peaks by ER convexities, and contact deflection reduces the maximum value of internal fluid pressure and makes the fluid peak sparse.

- The contact deflection is the only factor that generates the external fluid pressure, and only when the deflection of the ER gradually increases, the external oil film stress can be generated.
- The amplitude of the fluid force of ERSFD increases gradually with the eccentricity ratio. However, in the case of large eccentricity ratio, the ERSFD oil film force considering only convexity separation effect will have a large deviation.

Acknowledgements This study was supported by a grant from the national science foundation of China through Grant Nos. 12172088 and 51635004.

References

1. Andres LS (2014) Force coefficients for a large clearance open ends squeeze film damper with a central feed groove: experiments and predictions. Tribol Int 71:17–25
2. Chen WT, Chen SY, Hu ZH, Tang JY, Li HN (2019) A novel dynamic model for the spiral bevel gear drive with elastic ring squeeze film dampers. Nonlinear Dyn 98(2):1081–1105
3. Han ZF, Ding Q, Zhang W (2019) Dynamical analysis of an elastic ring squeeze film damper-rotor system. Mech Mach Theory 131:406–419
4. Li Y, Liao M, Wang S (2020) Effect on vibrating attenuation by the fit of elastic ring squeeze film damper. J Vib Shock 39(11):232–238
5. Yang XM, Jiang BZ, Li Y, Zhao Q, Deng SR, Zhang WH, Cui YC (2022) Dynamic characteristics of elastic ring squeeze film damper coupled high-speed ball bearings. J Sound Vib 537:30
6. Tian ZY, Hu ZH, Tang JY, Chen SY, Kong XN, Wang ZW, Zhang JE, Ding H (2023) Dynamical modeling and experimental validation for squeeze film damper in bevel gears. Mech Syst Signal Process 193:18
7. Wang ZL, Liu ZS, Zhang GH (2019) Dynamic characteristics of elastic ring squeeze film damper. Ind Lubr Tribol 71(10):1144–1151
8. Zhang W, Ding Q (2015) Elastic ring deformation and pedestal contact status analysis of elastic ring squeeze film damper. J Sound Vib 346:314–327
9. Sun K, Luo Z, Li L, Liu JX, Wu FY (2022) Dynamic analysis of the variable stiffness support rotor system with elastic rings. Nonlinear Dyn 110(1):201–217

A Power Generation Using Multiple Input DC-DC Converter for Small Water Turbines

Trung Kien Vu and Xuan Thanh Pham

Abstract Renewable energy systems are increasingly being developed to meet the demand for electrical energy while minimizing their environmental impact. Daily domestic water supply is stored on the top floors of buildings. This water generates pressure that can be harnessed to produce small-scale electricity. Using small-scale water turbines with low power output, minimal environmental impact, and low infrastructure costs, connect to a voltage converter to generate useful electricity for small device applications. This paper introduces the proposed multi-input DC-DC converter with different voltage levels controlled by MPPT, and PI algorithms to optimize the power output from the water turbines, achieving high efficiency with a wide input range and stable output voltage. The system includes a boost converter at the input stage and a buck converter at the output stage. The boost converter generates a stable output voltage of 35 V, while the buck converter produces an output voltage of 12 V with a high current. At a load current of 1 A, it achieves an efficiency exceeding 90%, significantly outperforming traditional designs.

Keywords Renewable energy · Small water turbine · DC-DC converter · Multiple input · High efficiency

1 Introduction

The substantial rise in human living standards has led to a large and increasing global energy consumption in the past decades. The production of this energy from fossil fuels accelerates the release of greenhouse gases and exacerbates environmental contamination [1–4]. Consequently, there is a growing need for sustainable energy sources, leading to extensive research efforts aimed at identifying novel sources or improvement of existing technology [5]. A 15-m-high building, a common building height of 4–5 floors, was considered for the study, with enough water to run the small

T. K. Vu · X. T. Pham (✉)
School of Electrical and Electronic Engineering, Hanoi University of Industry, Hanoi 100000, Vietnam
e-mail: thanhpx@haui.edu.vn

S. Y. Liang (ed.), *Manufacturing Technologies*, Lecture Notes in Mechanical Engineering, https://doi.org/10.1007/978-3-031-95455-9_8

turbines. The water utilized for our everyday tasks is typically pumped and stored in the roof tanks. This water supply along the pipeline generates pressure that drives small turbines and produces power [6]. Research and design of a voltage conversion circuit are necessary, this energy source capable of supplying adequate power and voltage for practical application and storage. Utilizing renewable resources for electric power generation is one of the most innovative approaches to energy production [7]. Therefore, any significant or minor stride towards energy conservation is highly valuable. Certain scholars postulate the feasibility of producing energy by utilizing water from rooftop water tanks attached to buildings. Hydropower is an excellent power generation system and a leading source of sustainable energy [8, 9], which generates kilowatts of electricity that can be stored in batteries and converted into alternating current (AC) and is frequently utilized in high-rise buildings worldwide. Furthermore, these systems necessitate buildings to have a height of 50 m or more and entail substantial initial expenses. This is unsuitable for buildings of relatively low height as well. To mitigate the impact on water pressure and facilitate the general implementation of renewable energy technologies for low-rise buildings, scholars have lately initiated investigations into water turbines that are significantly smaller and well-suited for micro-power production [10–12]. These exceptionally small turbines, referred to as nano water turbines in this study, have the capacity to generate power ranging from 5 to 10 W, and their blade dimensions are far smaller. These small water turbines do not necessitate extensive construction, are far more cost-effective, and are rather easy to maintain. They can be strategically placed in compact areas without compromising the structural integrity of the building. Despite the best suitability of extremely small water turbines for large-scale water energy generation, this technology nevertheless encounters certain obstacles. An inherent obstacle is the limited power generation rate of individual water turbines. Practical applications require the connection of several of these water turbines in order to generate sufficient electricity [13, 14]. This necessitates a multi-input DC-DC converter capable of accommodating the electrical output from several small-scale turbines linked together as a grid. As shown in Fig. 1, this work introduces a concept for a multi-input DC-DC converter that can be utilized in a series of low-power small water turbines. Moreover, the proposed converter combines with a Maximum Power Point Tracking (MPPT) algorithm control to aggregate the turbine's output voltage to match the voltage requirements of the load and the battery. This potential voltage source can be harnessed for use in small systems such as charging batteries and providing power for devices such as sensors, lights, warnings, measurements, and IoT (Internet of Things). This paper is established as follows. Section 2 proposes the design of a DC-DC converter with multiple input turbines. Section 3 describes the experimental results. Finally, Sect. 4 summarizes the conclusions of this work.

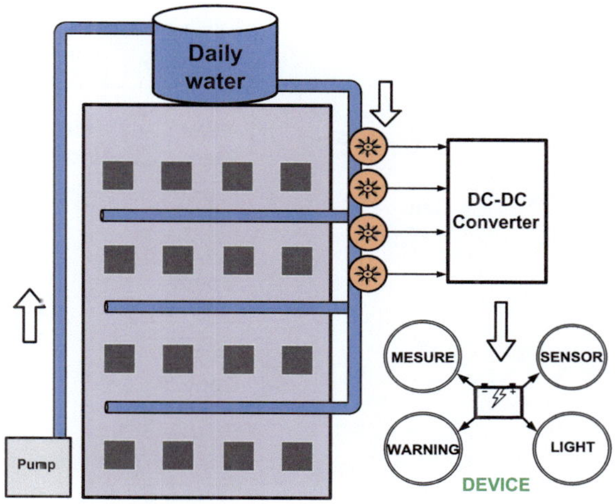

Fig. 1 The proposed circuit DC-DC converter model for small turbines used in buildings

2 Converter Design

Boost converters utilize a specific type of transistor, the N-channel MOSFET. This N-channel MOSFET has a lower R_{ON} resistance compared to the P-channel MOSFET, which helps achieve a higher ripple current in the inductor. The basic operating principle of a single-input boost converter involves using a clock pulse to switch on and off to increase the input voltage. However, this approach results in low efficiency and high output ripple. In the study [15], two parallel-connected boost converters with two inputs and 180-degree phase-shifted control pulses at a fixed 50% duty cycle were proposed to improve efficiency and reduce output ripple. However, the fixed duty cycle does not optimize the turbine's input power, as the water flow rate varies among different turbines in the system due to various factors, leading to energy losses. This paper proposes a dual-input boost converter with 180-degree phase-shifted control pulses and an adaptive duty cycle based on the MPPT algorithm to optimize the turbine's input power. The proposed converter design consists of two stages, namely, the input stage and the output stage. Figure 2 shows the schematic representation of the proposed design. In the first stage, the multi-input cascaded converter collects the voltage generated by each individual small-water turbine. This phase of the converter is designed with a modular architecture that incorporates several modular converters. This design can extend to integrate many small-water turbines and maximize the power generation capacity of the system. Assume that in the system, the input stage includes two boost converters A and B. By connecting the outputs of these cells in series, a series of DC-DC converters is formed, resulting in the addition of voltage.

Figure 3 illustrates the fundamental ideal of capacitor stacking, where two capacitors are charged and then connected in series. We can determine from KVL (Kirchoff's Voltage Law) that the voltage across the two capacitors is equal to the sum of the voltage on each capacitor. The first stage of this process generates a

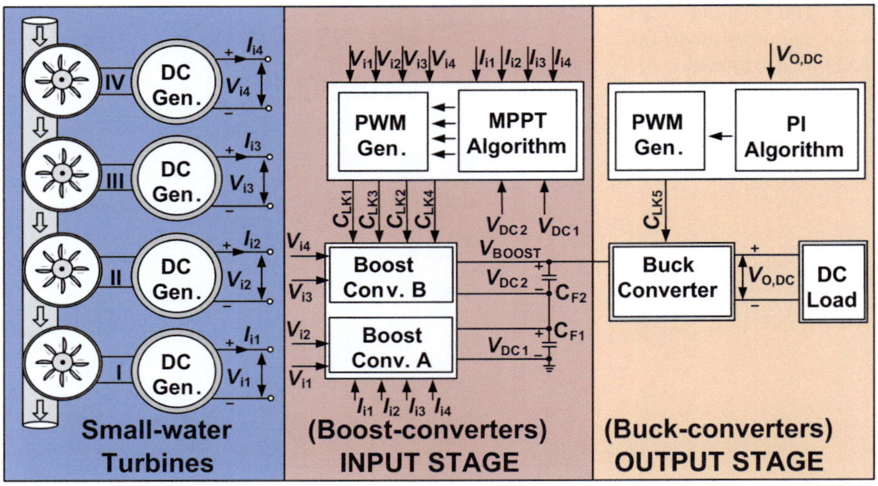

Fig. 2 Circuit diagram of multiple-input DC-DC converter with small-water turbines

voltage VBOOST that can be determined as $V_{\text{BOOST}} = V_{\text{DC1}} + V_{\text{DC2}}$ where $V_{\text{DC1,2}}$ is the DC output voltage of the boost converter A, and B, respectively. The purpose of stacking is to provide a high voltage input to the buck converter, which steps down the voltage to 12.6 V a heavy load current that can be stored and utilized. The output stage, the Buck converter of the system, regulates the output DC voltage $V_{\text{O,DC}}$ with high current to charge the batteries. Figure 4 shows the circuit for each boost converter in the proposed multi-input converter. Each turbine is connected to a separate boost converter circuit. The proposed architecture consists of two boost converters connected in parallel. Each converter uses an N-channel transistor with a separate clock control by the MPPT and PWM generation blocks, as shown in Fig. 4, to form a single boost converter block with multi-input. To improve efficiency, the N-channel transistor is controlled by an adaptive on-time cycle, which depends on the water flow rate of each turbine.

Additionally, the control frequency is set to 50 kHz to reduce ripple at the output (V_{BOOST}) of the input stage. As shown in Fig. 4a, during the positive half-clock cycle, in phase 1, the diode is reverse biased leading to D_2 being turned off and transistor Q_2 is turned on. while in phase 2, the diode is forward biased, causing D_1 to be turned on and transistor Q_1 turned off. The current I_{L2}(t) cannot pass through across

Fig. 3 The stacked structure is applied in the converter

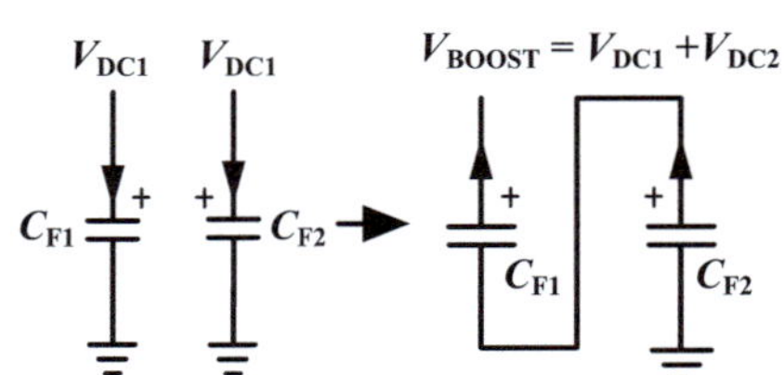

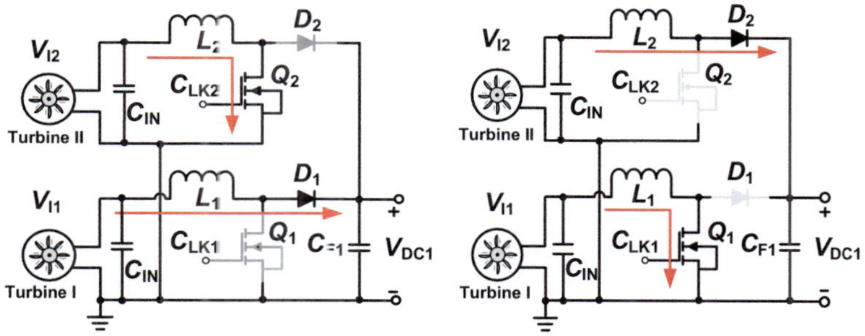

Fig. 4 **a** Positive half-clock cycle; **b** negative half-clock cycle of boost converter A and B

the diode. The current will charge the inductor (L_2), maintaining the energy in the magnetic form. When the switching cycle time is (TS), the input voltage of turbine II (V_{I2}) is used to calculate the inductor current ripple $\Delta I_{L2}(t)$, which is expressed by (1):

$$\begin{aligned} &0 < t < DT \\ V_{I2} &= L_2 \cdot \frac{d_{IL2}(t)}{dt} \\ I_{L2}(t) &= \frac{V_{I2}}{L_2} t + C \\ \Delta I_{L2}(\text{on}) &= \frac{V_{I2}}{L_2} DT \end{aligned} \tag{1}$$

As a result, the current is charging the capacitor C_{F1}. In Fig. 4b, during the negative half-clock cycle, in phase 1, the diode is forward biased leading to turn on D_2 and transistor Q_2 is turned off, while in phase 2, the diode D_1 is turned off and transistor Q_1 is turned on. Consequently, capacitor C_{F1} is still being charged. The current $I_{L2}(t)$ flows through the diode. The inductor will release the energy calculated shown in (2):

$$\begin{aligned} &DT < t < T \\ V_{I2} - V_{DC1} &= L_2 \cdot \frac{d_{IL2}(t)}{dt} \\ I_{L2}(t) &= \frac{V_{I2} - V_{DC1}}{L_2} t + C \\ \Delta I_{L2}(\text{off}) &= \frac{V_{I2} - V_{DC1}}{L_2} (1 - D)T \end{aligned} \tag{2}$$

where values of V_{I1} and V_{I2} are different because of different turbines. Constant current at the two conditions is expressed by (3):

$$\Delta I_{L2}(\text{on}) + \Delta I_{L2}(\text{off}) = 0$$

$$\frac{V_{I2}}{L_2}DT + \frac{V_{I2} - V_{DC1}}{L_2}(1 - D)T = 0$$

$$V_{DC1} = \frac{1}{1 - D}V_{I2} \tag{3}$$

The idea of the initial stages involves two boost converters connected in parallel, providing significant benefits in terms of energy accumulation [15, 16]. This configuration is advantageous because if water turbine I shuts down or malfunctions, water turbine II will continue to serve as an energy source and always maintain the power source to charge the capacitor, and the same is applied for turbines III, IV. Figure 5 illustrates the control pulses with an adaptive duty cycle for the boost converter, with inputs from turbines I and II. The MPPT algorithm is activated when the water turbine generates a voltage greater than 8 V, at which point the algorithm determines the optimal duty cycle for control, as shown in Fig. 5a. If the turbine motor generates a voltage below 8 V, the control pulses a duty cycle of 50%, as shown in Fig. 5b, to ensure that the different control clock signals do not overlap each other. With this control method, the overall converter can achieve maximum efficiency and reduced output voltage ripple. The turbines are designed to be installed in a cascaded system as shown in Fig. 2. One major advantage of this cascaded converter architecture is that the voltage output from each turbine or cell is summed, and any differences in voltage levels do not affect system performance, as shown in Fig. 6. In case of failure, the system can bypass malfunctioning turbines or cells, allowing it to continue operating with a slightly increased voltage ripple while still maintaining the output voltage. The design also provides safety benefits during installation and maintenance by enabling the isolation of each cell and its connected turbines for safe handling. Small water turbines are installed in series with a spacing of no more than 50 cm. But in reality, connecting four such turbines results in a dynamic pressure drop due to energy dissipation through the turbines, with no recovery as mentioned in studies [17, 18]. Therefore, exploitation of the water flow is maximized by proposing the turbines to be connected in series from turbines I, II, III, to IV but with small turbines rotating at different speeds, generating different voltages depending on the water speed. An MPPT (Maximum Power Point Tracking) algorithm with a step size determined by the error between V_{DC1} and V_{DC2} and the reference voltage is applied to optimize the converter's efficiency based on the power and input voltage of the small turbine generators. There are four control pulses, C_{LK1}, C_{LK2}, C_{LK3}, and C_{LK4}, to control the switching cycles T_{ON} and T_{OFF} of the two standard Boost Converters, and the voltage is collectively accumulated in capacitors C_{F1} and C_{F2}. The output voltage V_{BOOST} is 35 V, with V_{DC1} and V_{DC2} each being 17.5 V. Table 1 shows the parameters of the components and the calculated values of the Multiple Input Converter (Input Stage).

Figure 7 shows the circuit diagram of the output stage, a buck converter is utilized to regulate the voltage from $V_{BOOST} = 35$ V to $V_{O,DC} = 12.6$ V, which is suitable for various applications. The output voltage of the output stage is stable due to the application of the C_{LK5} pulse feedback controller, achieving high efficiency over a

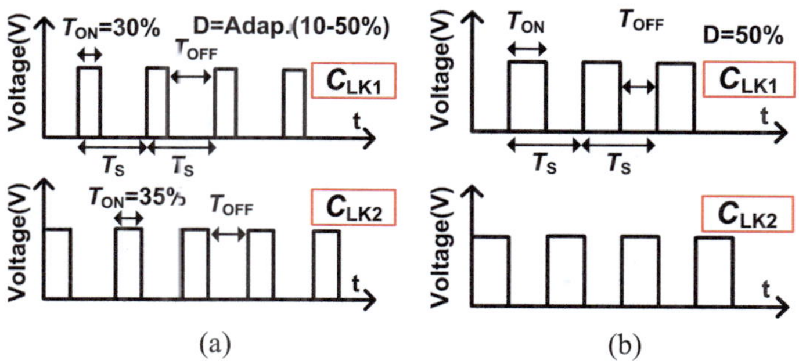

Fig. 5 The waveform of the proposed controller with: **a** the duty cycle with input voltage 5 from 8 V, **b** the duty cycle with input voltage 10 from 12 V

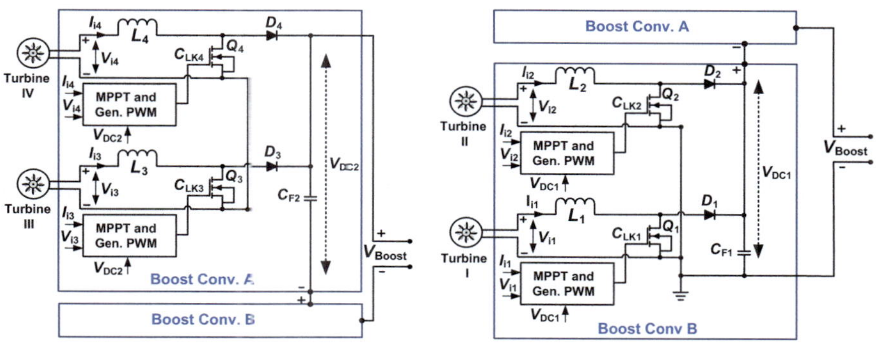

Fig. 6 The schematic of the input stage in the proposed converter

Table 1 The specification of the input stage in the proposed converter

Component	Type/value	Symbol
NMOS	IRFZ44N	Q_1, Q_2, Q_3, Q_4
Diodes	1N5408	D_1, D_2, D_3, D_4
Inductor	100 µH	L_1, L_2, L_3, L_4
Capacitor	1000 µF	C_{F1}, C_{F2}
Switches frequency	50 kHz	$C_{LK1}, C_{LK2}, C_{LK3}, C_{LK4}$

wide load range. To control the P-channel MOSFET (Q_5) with a high voltage on the drain, a BJT (Q_6) is used to assist the control pulses from the microcontroller with a low voltage threshold. Resistors R_{B1} and R_{B2} divide the voltage to limit the overload on Q_6. Table 2 contains the technical specifications of the components in the output stage, which have been optimally calculated. As shown in Fig. 8, the MPPT algorithm operates by continuously adjusting signals based on comparison methods, calculating current voltage and power values until the maximum power point is reached. The algorithm starts by comparing if $DP = 0$, meaning the power

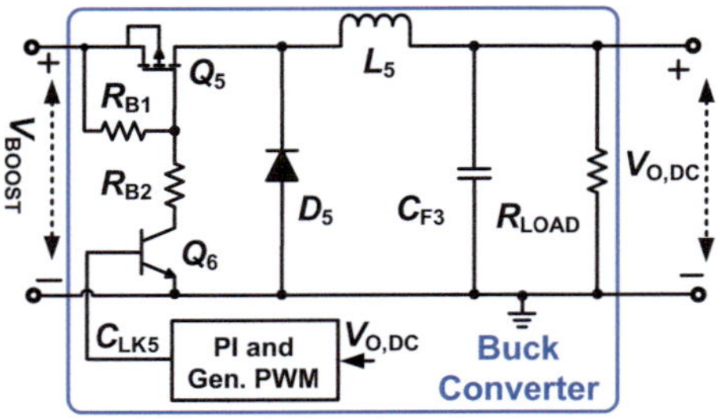

Fig. 7 The internal circuit diagram output stage in the proposed converter

Table 2 The specification of the output stage in the proposed converter

Component	Type/value	Symbol
PMOS	IRF4950N	Q_5
Diodes	1N5408	D_5
BJT	C1815	Q_6
Inductor	100 μH	L_5
Capacitor	1000 μF	C_{F3}

value is unchanged, then the new duty cycle D_{NEW} will be equal to D_{FB} which is the feedback duty cycle of the MOSFET from the output of the boost converter. The D_{FB} allows the algorithm to quickly reach the power point and stabilize the output voltage better than existing MPPT algorithms. If $DP < 0$, meaning the subsequent power is less than the previous power, and if $DV < 0$, the point is on the left side of the maximum power point and the duty cycle needs to be increased. If $DP < 0$ and $DV > 0$, the point is on the right side of the maximum power point and the duty cycle needs to be decreased. If $DP > 0$ and $DV < 0$, the duty cycle needs to be increased because the power is currently less and on the left side of the maximum power point, and vice versa. In particular, D_{NEW} is limited to not exceeding 50%.

3 Experimental Results

In the laboratory, four water turbines are connected in series through water pipes, the distance between the turbines is 100 cm. Therefore, the voltage values obtained at the water turbines IV, III, II, and I are 12, 10, 8, and 6 V, respectively as shown in Fig. 9. By using four small hydroelectric turbines with low power output, the maximum output voltage reaches 12 V. However, due to the effect of water pressure, when the turbines are connected in series, the voltage across each decreases. In order to do the DC-DC converter operations, Turbine I must generate a minimum voltage over a value of 5 V with sufficient power to supply the microcontroller. The layout experimental, STM32F103C8T6 microcontroller is used to control the proposed DC-DC converter with input and output stages. As shown in Fig. 10, the complete circuit design of the converter consists of two sections with two microcontrollers. Microcontroller

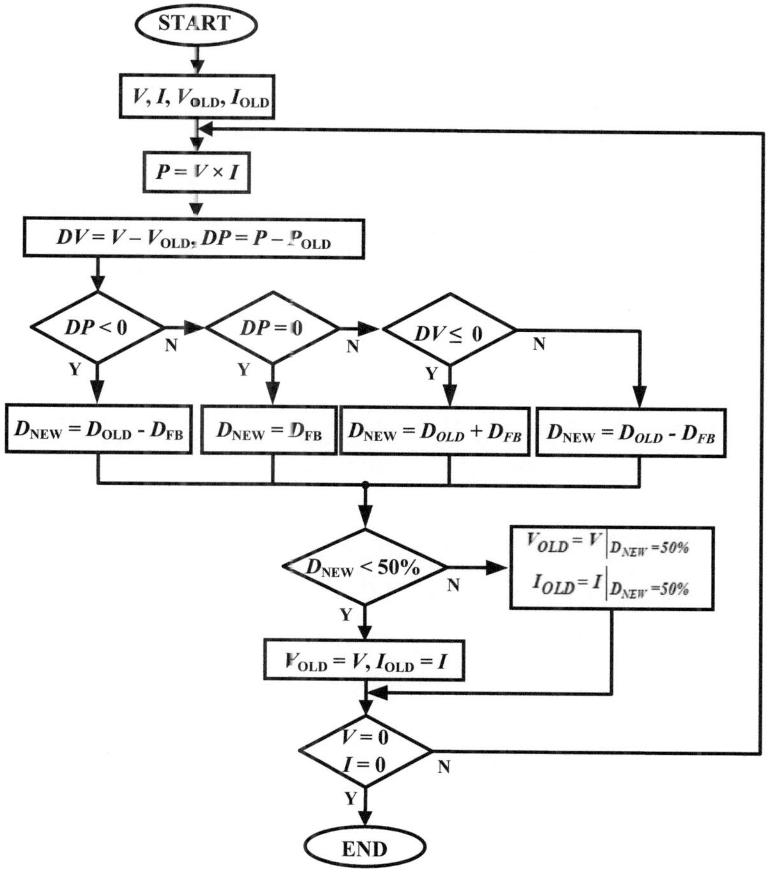

Fig. 8 The diagram MPPT algorithm for the system

1 generates pulses for boost converter A to control the voltage of turbines III and IV. Microcontroller 2 creates pulses for boost converter B to control the voltage of turbines I and II and control the pulse of the buck converter. The experiment consists of DC turbines connected coaxially to generate a DC power source for four inputs. According to Fig. 11, the Boost Converter A has control signals, C_{LK1} and C_{LK2} with a 180-degree phase shift. Due to the water speeds at turbines I and II being lower, the duty cycles of C_{LK1} and C_{LK2} are set at 50% to prevent efficiency loss. As shown in Fig. 12, the Boost Converter B has control signals C_{LK3} and C_{LK4} with adaptive duty cycles according to the speed of turbines III and IV, which are approximately 40%, and 33%, respectively. The output voltage of Boost Converters A and B are stable with $V_{DC1} = 17.5$ V and $V_{DC2} = 17.6$ V. By applying pulse controllers using the MPPT algorithm, the output voltages of Boost Converter A, and B are defined as stable. The duty cycles of the two Boost Converters are different due to variations in input voltages. As shown in Fig. 13, the voltage $V_{BOOST} = 35$ V is achieved due

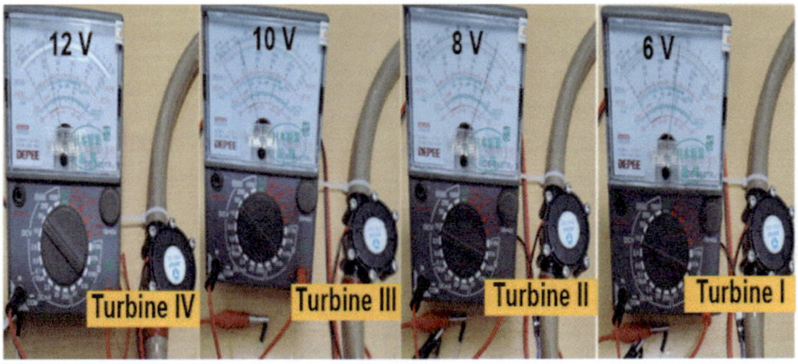

Fig. 9 The water turbines used in the proposed system

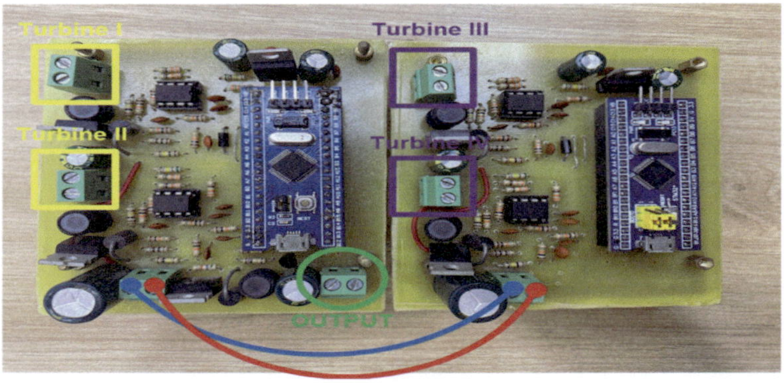

Fig. 10 The layout of the experimental systems

to capacitors C_{F1} and C_{F2} being stacked in series, effectively doubling the voltage. Finally, a high-efficiency Buck Converter provides a stable output voltage $V_{DC} =$ 12 V. Figure 14 shows the overall efficiency of the potential energy to electrical energy converter with varying load from 10 Ω to 1 kΩ. The efficiency of the proposed converter achieves an overall high 85% with a maximum of 90% at low load.

4 Conclusions

The paper presents the proposed DC-DC voltage converter suitable for multi-input small turbines installed in series within the water pipes of residential and commercial buildings to harness the daily domestic water supply stored on the top floors of buildings. By using small-scale water turbines with low power output, minimal environmental impact, and low infrastructure costs, connect to a voltage converter to generate useful electricity for small device applications. Moreover, the proposed DC-DC converter is controlled by MPPT, and PI algorithms to optimize the power

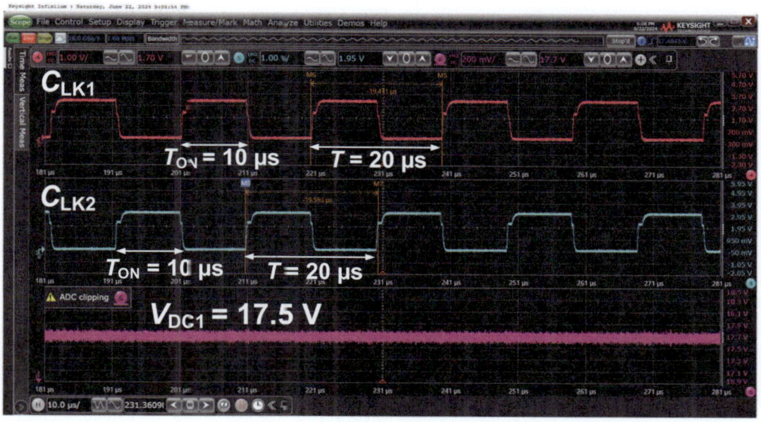

Fig. 11 The measured results of the signal C_{LK1} and C_{LK2} experimental

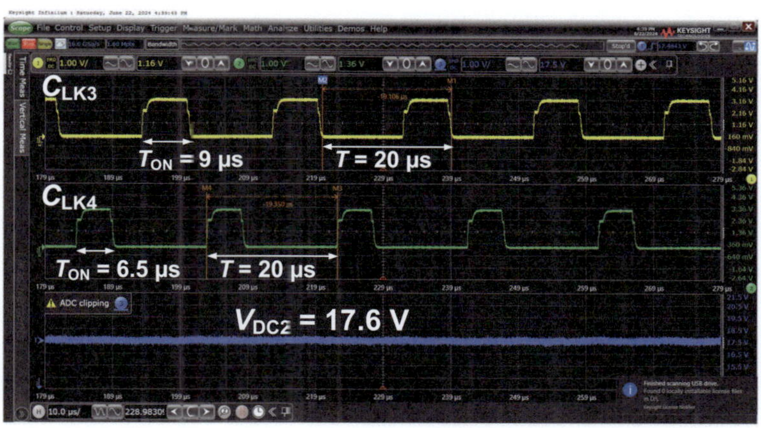

Fig. 12 The measured results of the signal C_{LK3} and C_{LK4} experimental

output from the water turbines, achieving high efficiency with a wide input range and stable output voltage. The system has been successfully implemented in a laboratory setting, demonstrating remarkable efficiency improvements. At a load current of 1 A, the design achieves an efficiency exceeding 90% with the output voltage of input and output stages of 35 and 12.6 V, respectively. This shows that the proposed design is suitable for the generated electricity from the small turbine applications and competitive with previous designs. In the future, the system can integrate more advanced controllers and more complex structures to enhance conversion efficiency and expand applications in various systems. Research and development of systems utilizing renewable energy sources is a global trend pursued by scientists, aiming to promote the growth of a green world.

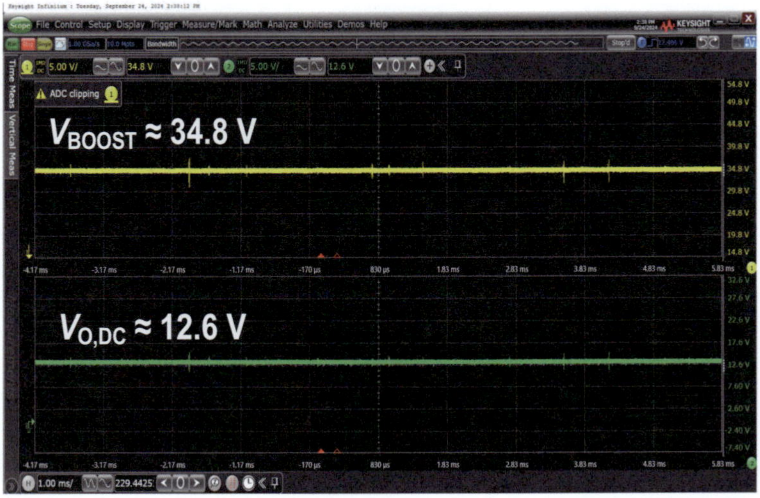

Fig. 13 The measured results of the voltage V_{BOOST} and $V_{O,DC}$ experimental

Fig. 14 The experimental efficiency of the DC-DC converter depends on the load

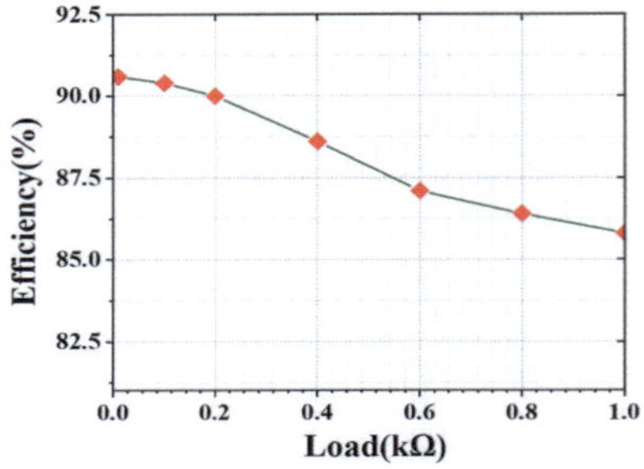

Acknowledgements The authors would like to acknowledge the support for this project from the Hanoi University of Industry Research Fund under grant number 27-2023-RD/HD-DHCN.

References

1. Mannan M, Al-Ghamdi SG (2022) Water consumption and environmental impact of multifamily residential buildings: a life cycle assessment study. Buildings 12(1):48
2. Daniel MT, Enjeti PN (2016) A power sharing scheme for series connected offshore wind turbines in a medium voltage DC collection grid. In: 2016 IEEE applied power electronics conference and exposition (APEC), Long Beach, CA, USA, pp 1695–1701
3. Bhati N, Kalla UK (2021) Investigations on PMBLDCG for micro hydro power generation. In: 2021 international conference on sustainable energy and future electric transportation (SEFET), Hyderabad, India, pp 1–4

4. Hashim N, Salam Z, Johari D, Ismail NFN (2018) DC-DC Boost converter design for fast and accurate MPPT algorithms in stand-alone photovoltaic system. Int J Power Electron Drive Syst (IJPEDS) 9(3):1038–1050
5. Shahzad U (2012) The need for renewable energy sources. Int Technol Electr Eng 2:16–18
6. Wang DA, Liu NZ (2011) A shear mode piezoelectric energy harvester based on a pressurized water flow. Sens Actuators, A 167(2):449–458
7. Nasir BA (2013) Design of micro-hydro-electric power station. Int J Eng Adv Technol 2(5):39–47
8. Carter J, Rahmani A, Dibaj M, Akrami M (2023) Rainwater energy harvesting using micro-turbines in downpipes. Energies 16:1660
9. Ge B, Xin P, Lv Y (2010) The excitation system simulation of huge hydro-generator. In: Asia-Pacific power and energy engineering conference, Chengdu, China, pp 1–4
10. Balkhair KS, Rahman KU (2017) Sustainable and economical small-scale and low-head hydropower generation: a promising alternative potential solution for energy generation at the local and regional scale. Appl Energy 188:378–391
11. Borkowski D (2018) Analytical model of small hydropower plant working at variable speed. IEEE Trans Energy Convers 33:1886–1894
12. Raza A, Xu D, Mian MS, Ahmed J (2013) A micro hydro power plant for distributed generation using municipal water waste with Archimedes screw INMIC. Lahore, Pakistan, pp 66–71
13. Abana E et al (2022) Micro hydro generator turbine. In: 14th international conference on computational intelligence and communication networks (CICN), Al-Khobar, Saudi Arabia, pp 672–676
14. Johnson MH, Aliprantis DC (2017) Analysis of series-dc offshore wind plants with aerodynamic wake effects. IEEE Trans Sustain Energy 8:1706–1714
15. Nur Ikhsan RR, Adib Arif K, Ramdhani M, Purnama I (2022) Dual-input cascaded DC-DC converter for hydromechanical energy harvesting in a hydroponic system. In: 2022 international conference on radar, antenna, microwave, electronics, and telecommunications (ICRAMET), Bandung, Indonesia, pp 228–233
16. Majeed YE, Ahmad I, Habibi D (2019) A multiple-input cascaded dc-dc converter for very small wind turbines. IEEE Trans Ind Electron 66:4414–4423
17. Giustolisi O, Ugarelli RM, Berardi L, Laucelli DB, Simone A (2017) Strategies for the electric regulation of pressure control valves. J Hydroinformat 19:621–639
18. Fontana N, Giugni M, Portolano D (2012) Losses reduction and energy production in water-distribution networks. J Water Resour Plann Manag 138:237–244

Image-Based Intelligent Control System and Monitoring

Communication-Driven Multi-Agent Path Planning Based on Improved Attention Mechanism

Yuhang Sheng and Huayi Yin

Abstract Multi-Agent Path Finding (MAPF) is crucial for achieving efficient coordination in large-scale robotic systems, particularly within the context of industrial intelligent decision-making and smart manufacturing. Although progress has been made in exploring decentralized policies using reinforcement learning (RL) in partially observable environments, developing collision-free strategies in dense environments remains challenging. This paper proposes a novel approach that integrates communication with deep Q-learning, enhancing inter-agent collaboration through graph convolution techniques. Unlike traditional methods, this study introduces an improved Efficient Channel Attention (ECA) mechanism to dynamically optimize channel responses within the network, significantly improving the model's ability to respond to critical environmental features. Furthermore, a position-encoding-enhanced multi-head attention mechanism is employed to optimize information exchange among agents, thereby improving decision efficiency and accelerating learning speed. The model is trained independently from the perspective of a single agent, and the learned policies are subsequently applied to each agent, ensuring decentralized execution. System-level training adopts a distributed architecture optimized via curriculum learning strategies. In densely obstructed testing environments, our approach demonstrates not only a high success rate but also reduced average steps, contributing to more effective intelligent decision-making processes in industrial applications.

Keywords Multi agent · Path planning · Efficiency channel attention · Deep Q-learning · Communication optimization

---

Y. Sheng · H. Yin (✉)
Xiamen University of Technology, Xiamen City, Fujian Province, China
e-mail: huayi@xmut.edu.cn

S. Y. Liang (ed.), *Manufacturing Technologies*, Lecture Notes in Mechanical Engineering, https://doi.org/10.1007/978-3-031-95455-9_9

1 Introduction

Multi-Agent Path Finding (MAPF) [1] is a core problem in achieving efficient cooperation among agents in large-scale robotic systems. Ensuring that multiple agents can travel simultaneously along their respective paths without collisions presents a key challenge in various real-world applications, such as automated intelligent warehousing, autonomous driving, and swarm robotics. MAPF aims to find an optimal set of collision-free paths for multiple agents to reach their respective goals from their initial positions.

In the context of smart logistics, the rapid development of e-commerce has made automated intelligent warehousing systems crucial. Leading companies like Amazon, Ocado, Fetch Robotics, Cainiao, JD Cloud Warehouse, and Geekplus have made significant advancements in this area. These systems require path planning for numerous agents to ensure efficient and safe destination arrival. For example, in automated distribution centers, warehouse robots autonomously navigate, reposition inventory, and deliver packages while following precise, collision-free paths.

This study aims to optimize inter-agent communication and improve decision quality by introducing an improved Efficient Channel Attention (ECA) mechanism and a position-encoding-enhanced multi-head attention mechanism. This novel approach enhances path planning efficiency, avoids collisions in congested environments, and optimizes time efficiency. Our model adopts a decentralized execution approach, where each agent makes independent decisions based on local observations, while system-level training occurs in a distributed framework utilizing curriculum learning [2] strategies. Extensive empirical testing demonstrates that our method outperforms existing techniques across multiple key performance metrics, validating its effectiveness in real-world scenarios.

Our main contributions are as follows:

1. Introducing the improved Efficient Channel Attention (ECA) [3] mechanism to enhance the model's ability to identify key environmental features, enabling agents to execute tasks more accurately in complex environments;
2. Utilizing the position-encoding-enhanced multi-head attention mechanism to improve communication efficiency among agents, optimizing decision-making processes and accelerating learning speed;
3. Extensive empirical evaluation showing that our approach outperforms existing techniques in key metrics such as success rate and average number of steps.

2 Problem Definition

This paper studies the multi-agent pathfinding (MAPF) problem, which is formalized using the following model: Consider an undirected graph, and a set of agents. Each agent has a designated start vertex and goal vertex. At each discrete time step, an agent can choose to move to an adjacent vertex or remain at its current vertex.

The path of an agent is represented by a sequence of vertices, starting from the start vertex and ending at the goal vertex. Consecutive vertices in the sequence represent the agent's movement, while repeated vertices indicate waiting. Collisions between agents can occur as either vertex collisions or edge collisions: A vertex collision is defined by the quadruple, which occurs if and only if agents and reach the same vertex at the same time. An edge collision is defined by the quintuple, which occurs if agents traverse the same edge in opposite directions at time.

The objective of solving the MAPF problem is to find a collision-free path for each agent. The quality of the solution is measured by the total time for all agents to reach their goal vertices. This study focuses on a two-dimensional four-neighbor grid model, while the proposed approach can be extended to higher-dimensional grid models.

3 Related Work

3.1 Environment Settings

To meet the requirements of standard Multi-Agent Pathfinding (MAPF) tasks, this study focuses on a two-dimensional four-connected grid, where agents, goal positions, and obstacles each occupy an independent grid cell. Specifically, the map is defined as an m $\times$ m matrix, with "0" representing a free position and "1" representing an obstacle.

On the map, n agents are randomly assigned n starting positions and corresponding n goal positions, all of which are free cells. We ensure that each goal position is reachable from its respective starting point, and there is no overlap among the selected 2n positions. At each time step, all agents can simultaneously move to adjacent positions or wait in their current position, resulting in an action space of size 5 (four directional movements plus staying in place). During the simulation, agents may encounter obstacles or collide with other agents. These collisions are handled by recursively resetting the affected agents to their previous states until no conflicts remain.

Furthermore, considering that partially observable environments better reflect real-world scenarios, each agent can only obtain information within its Field of View (FOV). The FOV is set to a size of k $\times$ k (where k is an odd number to ensure the agent is at the center of the FOV), and k is significantly smaller than m.

3.2 Reward Design

Regarding the reward mechanism, this study adopts the design from DHC [4], as detailed in Table 1. Typically, to encourage agents to reach their target locations as

Table 1 Reward function design

Actions	Reward
Move (up/down/left/right)	−0.075
Stay (on goal, away goal)	0, −0.075
Collision (obstacle/agents)	−0.5
Finish	3

quickly as possible, a negative reward is applied for each movement. Unlike PRIMAL [5] and MAPPER [6], where agents receive a larger penalty for staying still, we assign the same penalty for both moving and staying in place (if not at the goal), since in more complex scenarios, an agent might need to stop to allow another agent to pass and avoid collisions.

4 Methods

4.1 Improved Distributed Heuristic Learning with Communication

In this section, we provide a comprehensive design of our model, referred to as DHC-ECA, which encompasses three key components: Distributed, Heuristic, and Communication, along with the integration of the ECA Mechanism. The primary objectives of DHC-ECA are to leverage heuristic guidance in reinforcement learning (RL) and to facilitate agent cooperation through independent learning paired with graph convolutional communication. This independent learning framework can be effectively distributed, which helps to accelerate the learning process and improve overall performance.

We construct a Dueling DQN [7] with recurrent units to approximate an agent's policy in partially observable gridworlds. This model is designed to map current observations and communication messages to the Q-values of each action, featuring an enhanced agent network composed of three modules: observation encoder, communication block, and Q-network.

The observation encoder now consists of eight convolutional layers and a GRU [8]. To enhance feature extraction and highlight important channels, we have integrated the Efficient Channel Attention (ECA) mechanism into each residual block of the convolutional layers. This attention module helps the model focus on significant features, thereby enhancing its representation capabilities without excessive computational cost. The convolutional layers are organized as three residual blocks, each incorporating the ECA module, along with two additional independent convolutional layers. For the agent network, the local observation of agent i is encoded into $\hat{o}_i^t$ by the convolutional layers enriched with ECA. Subsequently, a GRU processes

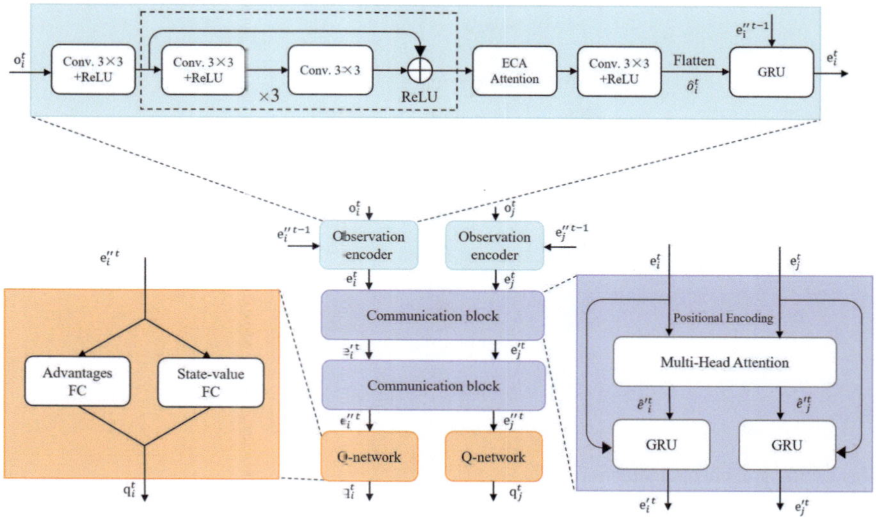

Fig. 1 DHC-ECA consists of three modules

$\hat{o}_i^t$ alongside the hidden state to generate the intermediate message $e_i'^t$. This message incorporates important historical information for the current step communication.

In addition, we have introduced positional encoding within the communication block to enhance the understanding of spatial relationships among agents. This is particularly crucial for complex multi-agent interactions, where the positional information can aid in accurately estimating the distance and influence among agents. The communication block consists of a graph convolutional network followed by a GRU. Messages e_i^t and $e_{\mathcal{B}_i}^t$, where $\mathcal{B}_i$ represents the neighbors of agent i, are integrated through graph convolution to produce a feature vector $\hat{e}_i^t$. Positional encoding is applied to the query and key components within the attention mechanism, ensuring that spatial positions are taken into account during communication. This enriched feature vector $\hat{e}_i^t$ is then passed to the GRU, along with e_i^t serving as the hidden state, to generate information $e_i'^t$. Through multiple communication rounds, agents gradually gather more meaningful information from their neighbors, leading to a broader cooperative scope. The output from the final communication round, $e_i''^t$, is used by the Q-network to estimate Q-values in a dueling fashion, where both the state value and action advantage are calculated and combined to obtain the optimal policy, see Fig. 1.

4.2 *Efficient Channel Attention*

The Efficient Channel Attention (ECA) module is designed to improve the model's ability to identify and focus on key features by dynamically adjusting the significance

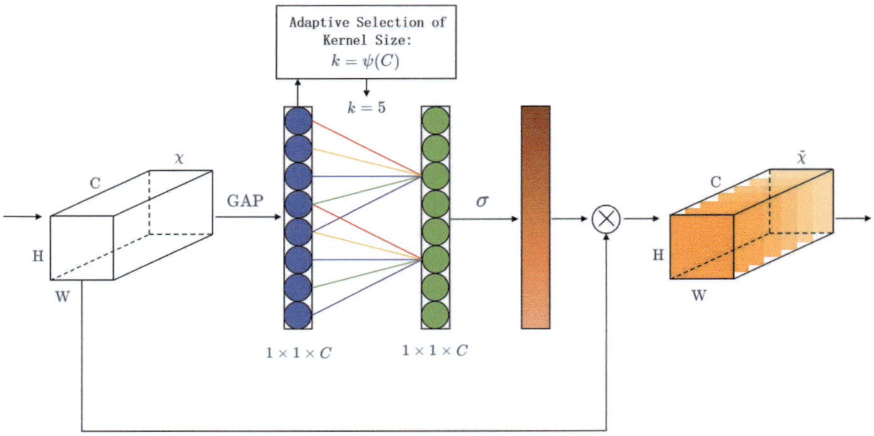

Fig. 2 Efficient channel attention

of each channel. Unlike conventional attention mechanisms that require explicit dimensional transformations, such as downscaling and upscaling, ECA eliminates the need for these operations, thus reducing both the parameter count and computational complexity. This makes ECA particularly advantageous in scenarios where computational resources are limited, as it maintains the effectiveness of traditional attention mechanisms while being more efficient. By applying a lightweight 1D convolution, ECA is capable of capturing channel-wise dependencies effectively, allowing the model to adaptively emphasize channels that carry more informative features, see Fig. 2.

In the context of multi-agent systems, ECA enhances the network's sensitivity to the environment by accurately weighing the importance of various features across channels. This increased sensitivity helps the model better understand and predict complex interactions among agents, ultimately leading to more informed decision-making. The improved feature representation, combined with reduced computational overhead, makes ECA highly suitable for real-time applications where quick and effective responses are crucial. Overall, the integration of ECA enhances the adaptability and performance of the model, making it a powerful addition to complex multi-agent pathfinding tasks.

4.3 Introducing Efficient Channel Attention and Position Encoding

The introduction of the Efficient Channel Attention (ECA) module offers multiple advantages. By dynamically assigning weights to features across channels, the

ECA module enhances the model's capacity to adaptively determine the significance of each channel. This improvement enables the network to better differentiate between channels that carry rich information and those that do not, thereby enhancing its ability to capture critical features. Unlike traditional attention mechanisms, ECA does not require explicit dimensional transformations (such as downscaling and upscaling), which significantly reduces parameter count and computational complexity. This makes the ECA module especially well-suited for resource-constrained environments, while preserving the benefits of attention mechanisms. In multi-agent systems, the ECA module significantly bolsters the model's sensitivity to the states of agents within the environment, enabling the network to more accurately interpret and predict complex interactions between agents, thereby enhancing overall adaptability. This directly enhances decision-making quality, resulting in higher task success rates and fewer required steps.

By integrating positional encoding into the attention mechanism, the model can capture both relative and absolute positional information between agents, which is critical for learning spatial strategies. Positional encoding adds valuable spatial context, making the model more effective and accurate in processing interactions between agents. In complex multi-agent environments, the relationships between agent positions often determine the nature and outcome of their interactions. Positional encoding improves the model's understanding of these relationships, enabling decisions that align more closely with the actual spatial layout, particularly in complex or dynamically changing environments. Without positional encoding, the model often struggles to capture long-range dependencies between agents. Introducing positional encoding allows the model to effectively encode and utilize long-distance spatial information, which is essential for executing complex, multi-step tasks or responding in dynamic environments.

Incorporating both the ECA module and positional encoding into the multi-agent system not only improves the efficiency and effectiveness of the attention mechanism but also significantly enhances the model's grasp of spatial dynamics in complex multi-agent environments. These advancements provide a practical means to boost system performance and are expected to contribute substantially to real-world applications, which can be highlighted in the paper.

5 Experimental Design and Evaluation

We evaluated our model's performance by comparing its success rate and average step count against the state-of-the-art DHC and the established PRIMAL approach. The success rate measures the model's ability to complete tasks within a specified number of time steps, while the average steps indicate the mean time required to complete these tasks, with a lower value reflecting a more optimal strategy. Experiments were conducted in an environment identical to that used by DHC to ensure a fair and rigorous comparison, considering two map sizes: 40×40 and 80×80. To emphasize the ability of DHC-ECA to manage complex environments, the obstacle density was

Table 2 Average Step in two different environments

Average step	Map size 40 × 40			Map size 80 × 80		
Agents	DHC-ECA	DHC	PRIMAL	DHC-ECA	DHC	PRIMAL
4	**50.16**	52.33	79.08	**92.95**	96.72	134.86
8	**56.00**	63.90	76.53	**105.43**	109.24	153.20
16	**67.66**	79.63	107.14	**116.22**	122.54	180.74
32	**86.92**	100.10	155.21	**127.65**	138.32	250.07
64	168.52	**147.26**	170.48	**149.69**	163.50	321.63

set at 0.3—the highest level used by both DHC and PRIMAL. For each agent count {4, 8, 16, 32, 64}, we conducted 200 test cases, with a maximum of 256 time steps for the 40 × 40 map and 386 for the 80 × 80 map. The results presented in the figures show that our method consistently outperformed DHC and PRIMAL across all scenarios.

As indicated in Table 2, in congested environments, such as scenarios involving 32 and 64 agents on the 40 × 40 map, DHC-ECA exhibited mixed performance. Specifically, for 64 agents, DHC-ECA performed slightly worse than DHC in terms of average steps, though it still surpassed PRIMAL. This outcome suggests that DHC may hold a slight advantage in managing collisions in highly congested conditions. However, in other settings, DHC-ECA's average step counts were significantly lower than those of both DHC and PRIMAL, demonstrating the benefits of dynamic channel response optimization in enhancing responsiveness to critical features, thereby improving coordination and overall success rate.

Figures 3 and 4 further validate the advantages of DHC-ECA. Notably, when transitioning from the 40 × 40 to the 80 × 80 map, the increase in average steps for DHC-ECA was minimal, indicating its ability to effectively learn from heuristics and identify efficient paths in larger environments. Furthermore, DHC-ECA consistently required fewer time steps to complete tasks, underscoring the benefits of enhanced communication in managing collisions and expediting pathfinding. These findings strongly suggest that DHC-ECA's approach to integrating dynamic channel attention can significantly improve performance in multi-agent navigation tasks, particularly in complex and resource-constrained environments.

6 Conclusion and Prospect

We propose an improved attention-based, communication-driven multi-agent pathfinding method, introducing the Efficient Channel Attention (ECA) mechanism to the multi-agent navigation problem. This mechanism dynamically adjusts the channel response in the network, enhancing the model's ability to identify key environmental features. Our approach combines multi-head attention with positional encoding, significantly improving coordination efficiency among agents. Inspired by

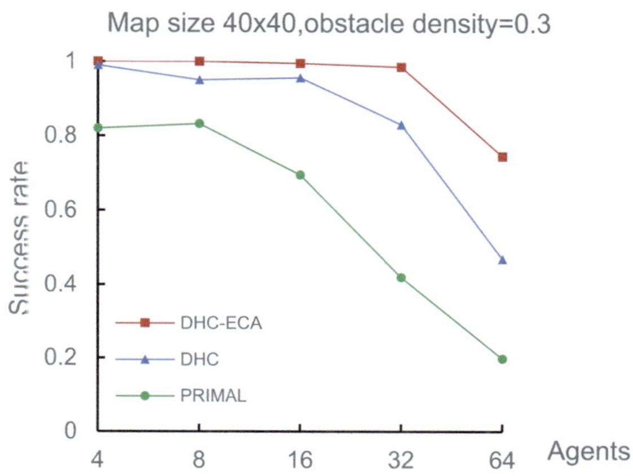

Fig. 3 Success rate of our method compared with DHC and PRIMAL, Map size 40 × 40

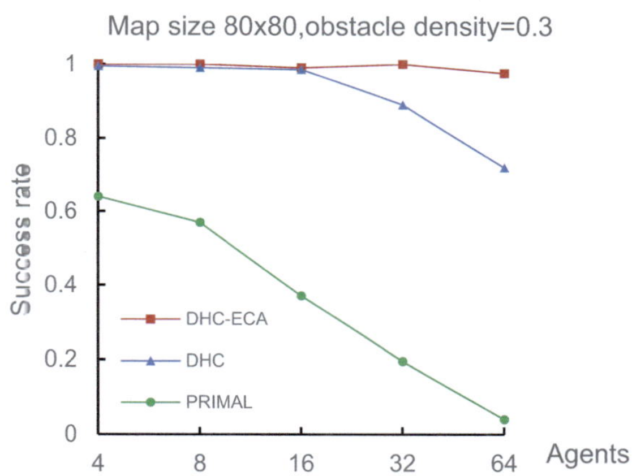

Fig. 4 Success rate of our method compared with DHC and PRIMAL, Map size 80 × 80

DHC, we employ a single-source shortest path embedding in reinforcement learning, instead of using a specific single-agent shortest path. Graph convolution is used for communication, enabling effective collaboration between agents. The model is trained in a distributed manner from the perspective of individual agents, ensuring excellent scalability.

Experimental results show that compared to DHC and PRIMAL, the proposed DHC-ECA, with enhanced channel attention, achieves the shortest average step length in most cases for both sparse and dense environments, and the highest success rate across all situations. Our method significantly improves agent coordination and navigation efficiency in complex scenarios, contributing to advancements in industrial intelligent decision-making and smart manufacturing applications. In the future, we aim to further optimize the dynamic channel response to enhance the model's ability to capture key environmental features, thereby improving performance. We also plan to explore crowded and complex environments further, with a focus on applications in industrial automation and intelligent logistics.

Acknowledgements This work is supported in part by Natural Science Foundation of China (No. 61773325), Natural Science Foundation of Fujian Province (Nos. 2021J011182, 2021J011191), Xiamen Science and Technology Plan of University Innovation Project (Nos. 2022CXY0401, 2024CXY0432).

References

1. Stern R, Sturtevant N, Felner A, Koenig S, Ma H, Walker T, Li J, Atzmon D, Cohen L, Kumar T et al (2019) Multi-agent pathfinding: definitions, variants, and benchmarks. In: Symposium on combinatorial search, pp 151–159
2. Bengio Y et al (2009) Curriculum learning. In: Proceedings of the 26th annual international conference on machine learning
3. Wang Q et al (2020) ECA-net: efficient channel attention for deep convolutional neural networks. In: Proceedings of the IEEE/CVF conference on computer vision and pattern recognition
4. Ma Z, Luo Y, Ma H (2021) Distributed heuristic multi-agent path finding with communication. In: 2021 IEEE international conference on robotics and automation (ICRA), Xi'an, China, pp 8699–8705
5. Sartoretti G et al (2019) PRIMAL: pathfinding via reinforcement and imitation multi-agent learning. IEEE Robot Autom Lett 4(3):2378–2385
6. Liu Z, Chen B, Zhou H, Koushik G, Hebert M, Zhao D (2020) Mapper: multi-agent path planning with evolutionary reinforcement learning in mixed dynamic environments. In: Proceedings of the IEEE international conference on intelligent robots and systems (IROS'20)
7. Wang Z, Schaul T, Hessel M, Hasselt H, Lanctot M, Freitas N (2016) Dueling network architectures for deep reinforcement learning. In: Proceedings of the international conference on machine learning (ICML'16), pp 1995–2003
8. Chung J et al (2014) Empirical evaluation of gated recurrent neural networks on sequence modeling. arxiv:1412.3555

The Effect of Acquisition Time on Thermographic Monitoring and Identification of LRM Process Dynamics

Patrick R. Meyer, Evgueni V. Bordatchev, and O. Remus Tutunea-Fatan

Abstract This study investigates the effect of acquisition time on the quality and information of laser remelting process (LRM) dynamics by means of a near-infrared (NIR) camera. Experiments were conducted with varying laser powers (7.5, 10, 12.5 W) and acquisition times (50–200 μs). The results suggest that acquisition time constitutes a critical element of the control process since it affects the observability and identifiability of LRM dynamics. More specifically, optimal exposure time balances acquisition rate and quality, with 50 μs being the most effective value. Future research will explore strategies to improve data acquisition and develop closed-loop control systems for LRM processes.

Keywords Laser remelting · Laser polishing · Process dynamics · Thermographic monitoring · Acquisition time · Closed loop process control · Step response · Transfer function · Identification

1 Introduction

The laser remelting process (LRM) is a widely applicable surface topography modification technique with multidisciplinary uses, such as surface polishing and surface structuring due to its superior controllability, versatility, and ability to improve topographic and physical characteristics of the surfaces impacting functionality, geometry, and/or appearance [1, 2]. The LRM process uses laser radiation for rapid remelting and solidification of a small molten pool on the surface that alters form and

P. R. Meyer · O. Remus Tutunea-Fatan (✉)
Department of Mechanical and Materials Engineering, Western University, London, ON, Canada
e-mail: rtutunea@eng.uwo.ca

E. V. Bordatchev
Automotive and Surface Transportation, National Research Council of Canada, London, ON, Canada
e-mail: evgueni.bordatchev@nrc-cnrc.gc.ca

S. Y. Liang (ed.), *Manufacturing Technologies*, Lecture Notes in Mechanical Engineering, https://doi.org/10.1007/978-3-031-95455-9_10

topography [3]. Surface polishing by laser remelting (SP-LRM) denotes the application of LRM to produce an automated and noncontact surface quality improvement process using the melt pool's thermodynamics, such as: surface tension, capillary, and thermocapillary flows [4]. The performance of the laser remelting process is influenced by a complex combination of the selected process parameters and process dynamics such as power fluctuation, material property changes, nonuniformity of initial surface topography, and the laser path [5, 6]. This can lead to unstable conditions because of external disturbances or unknown instabilities making fixed process parameters insufficient for operation. Determining optimal parameters is difficult, therefore, to improve reliability, repeatability, and the overall surface quality, a closed-loop process control system is desirable to compensate for process disturbances.

Monitoring the system is important for control, but traditional methods of measuring surface temperature pose challenges due to process conditions and speed. This is addressed using a thermographic monitoring system as a noncontact data acquisition method which records thermal emissions and creates a thermographic representation that is correlated with temperature distribution [7, 8]. Other laser-based manufacturing processes such as laser powder bed fusion and laser alloying–have implemented real-time feedback from process sensors to integrate data for automation and control of process parameters and laser melt pool dynamics [9, 10]. In this case, process feedback is a measured signal corresponding to surface temperature in the laser-material interaction zone or as parameters of the thermographic image. Acoustic emissions can also be used for monitoring purposes [11].

The existing control strategies modeled the laser-material interaction process with first and second order linear dynamic systems [9, 10] that were reduced to a transfer function (TF) between input process parameters (laser power, scanning speed) and a parameter calculated from online measured temperature distribution, verifying that this method works for LRM is important considering the differences in operating power and dynamics between processes. As an alternative option, TF between initial and remelted surface topographies can be used, but this is only performed offline [6, 12]. First-order systems proved most effective for modelling and control, while second-order systems introduce both overshoot and undershoot reflects regions of rapid heating and immediate cooling. The transient behavior of second-order systems was not observed in the LRM trials and is unlikely to be appropriate for modelling.

The typical first order TF is comprised of two terms–a proportional gain, K, and a time constant, τ:

$$TF_{LRM}(s) = \frac{\text{parameter of measured temperature distribution } T(x, y, s)}{\text{process parameter } (W(s) \text{ or } v(s))} = \frac{K}{1 + \tau s}, \tag{1}$$

where s is the Laplace operator, W is the laser power, and v is the scanning speed.

The LRM process TF, TF_{LRM}, can be identified by measuring a step response function to be calculated when an instant change of W or v is applied and $T(x, y, t)$

is recorded. This study recorded $T(x, y, t)$ under different laser powers and acquisition times in order to explore the tradeoff between a minimum control cycle time (prerequisite of real-time process control implementations) and the ability to obtain instantaneous, reliable, and informed feedback about the LRM process dynamics. This work advances the scientific and engineering understanding of LRM process thermodynamics, enabling the development of physics-informed digital twins and AI-based process modeling, and control.

2 Automatic Control of the LRM Process

A typical automatic control theory-based closed-loop system can be adapted and implemented for online control of the LRM process (Fig. 1). This requires four main components; controller, model of the LRM thermodynamics, sensor(s), and data acquisition and processing unit (DAPU). The controller accepts and compares process related parameters (speed, power, etc.) and feedback variables from actual process performance, and uses a control algorithm to generate a control action (change laser power or scanning speed, etc.). The LRM thermodynamics is a complex nonlinear process typically requiring linearization, making it the most critical and difficult determine, and it is used in conjunction with the TF of the sensor to define the control algorithm. The sensor (thermographic camera or pyrometer) observes the physical laser-material interaction state and converts it into signal to compute feedback variables. The DAPU uses its input signal to determine parameters of the observed state (i.e. calculating area-at-level, volume-above-level, etc. from thermographic images). The computed parameters serve as feedback variables that close the control cycle. Various disturbances of the LRM process (variable absorptivity or preheating/overheating/underheating conditions) translate into TG image changes and alter process parameters sent to the controller. Once a control algorithm is defined the online LRM process can substitute the thermodynamic model to determine reactions in response to real-time perturbations.

Since the sensors measuring acoustic [11], optical [13], and electromagnetic [14] emissions are characterized by a very high acquisition rate (up to 100 kHz), the control cycle interval depends solely on the controller performance that also tends to be quite high. However, these sensors are unidirectional and point-oriented and measure an integrated response without information on the spatial dynamics of the molten pool.

Thermographic cameras overcome this challenge by capturing spatial thermographic emission using a complementary metal oxide semiconductor (CMOS) image sensor. The sensor consists of a matrix of miniature photodiodes and microlenses for each pixel which captures an element of the thermographic emission and assembles a spatial image of the laser process. The exposure is critical as it determines the period each photodiode accumulates voltage on the thermographic emission and produces an integrated response [15]. For this reason, exposure time is a key factor in obtaining the resulting image within the control cycle. Typically, thermographic cameras have

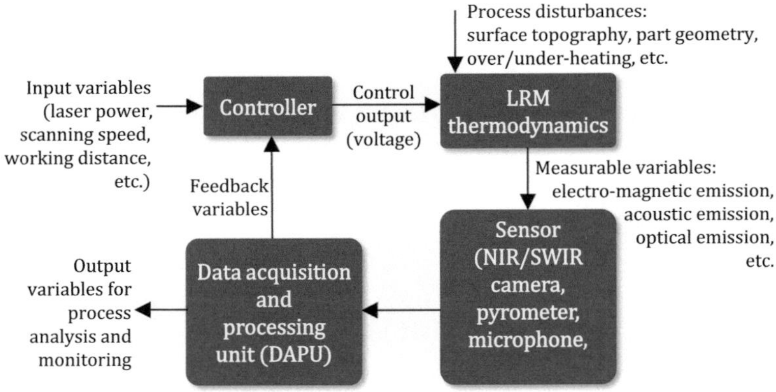

Fig. 1 Structure of the closed-loop system for automatic control of the LRM process

up to 100 fps for the entire image matrix, and reducing matrix dimensions can increase it to 10,000 fps.

The acquisition time defines the imaging speed but at the same time, it affects the information quality. More specifically, there is a critical tradeoff between shorter acquisition times for the control cycle interval resulting in less information being captured and larger acquisition times for more reliable information which can obscure short-time process variations to consider. The present study systematically analyses the effect of the acquisition time on the process informativity and reliability of the TF_{LRM} experimental identification for future applications.

3 Experimental Setup and Analysis Methodology

3.1 Experimental Setup

For experimental identification of TF_{LRM} an advanced experimental setup is needed. The setup used for this study integrates a high-acceleration laser scanner–for delivery of an instant SRF with respect to the process thermodynamics parameters (laser power, speed, etc.); a coaxially located thermographic camera with controllable acquisition time–used to acquire a thermographic response directly from the laser-material interaction zone, and perhaps most importantly, software for synchronous online data acquisition [5]. The experimental setup used in this study (Fig. 2) integrates several opto-electro-mechanical components and systems: (1) three-axis motion system from AEROTECH mounted on a granite table for precise positioning a workpiece within a work envelope; (2) two-axis laser scanner (exelliSCAN 14, SCANLAB GmbH, Germany) with a f-theta lens having a focusing distance of 256 mm; (3) continuous wave (cw) laser from IPG with a wavelength of 1070 μm and laser power up to 100 W delivered via an optical fiber into the laser scanner after collimating the laser beam; (4) optical adapter for mounting a thermographic camera and redirecting the thermal emission signal from the laser-material interaction zone into the camera; (5) coaxially mounted thermographic camera Nano M640-NIR from Teledyne Dalsa having an image resolution of 640 × 480 pixels and an 8-bit monochrome CMOS sensor with enhanced NIR sensitivity capable of acquiring images with 850 fps; and (6) CNC motion control system from AEROTECH for positioning linear drives with a mounted workpiece. An in-house developed software for online data acquisition was used for the synchronous recording of the actual time, real X-Y scanner positions, and thermographic 640 × 480 pixels with an acquisition rate of up to 6000 fps. A workpiece with 50 mm × 50 mm × 10 mm dimensions from ground H13 tool steel was mounted on X-Y linear stages within an Argon-filled enclosure to prevent oxidation caused by the atmospheric environment.

Since this study was only focused on the effect of exposure times, experiments consisted of 8 mm-long individual lines performed with a scanning speed of 50 mm/s and with three different laser powers: 7.5, 10, and 12.5 W. Experiments were performed such that the initial 4 mm of the line were with laser power switched off whereas the remaining 4 mm were generated with laser power on. Each experiment was repeated four times with four different exposure times: 50, 100, 150, and 200 μs.

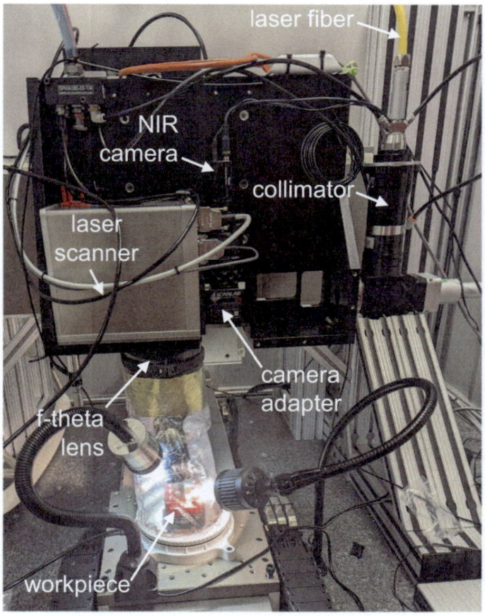

Fig. 2 Experimental setup and its components

3.2 Experimental Data Analysis Methodology

LRM dynamics can be regarded as a single-input single-output (SISO) continuous time system under linear approximation. The instant switch of the laser power at 0 mm mark (4 mm after the start of the line) constitutes an input step function response (approximation of Heaviside step function) and therefore TG images will represent a transient state continued as a steady-state process until the end of the laser track where laser power will be switched off. The sequence of recorded TG images during the transient state is considered as SRF_{LRM} as a response to the step input. Each thermographic image consists of a 64 × 64-pixel matrix and cannot be considered as a single output. Because of this, past studies have shown that many different parameters can be computed from each thermographic image [16].

Here, two computationally efficient parameters of the TG images were selected because of their anticipated applicability to the intelligent adaptive control of the LRM process. The first parameter is an area-at-level, A_{level}, calculated as a number of pixels with amplitudes equal or above a specified level in bits. The second parameter selected is a volume-above-level, V_{level}, calculated as a sum of TG image amplitudes for each pixel having an amplitude equal or above specified level in bits. Three levels were chosen for analysis: 10 bits, 100 bits, and 200 bits. The 10-bit level was chosen to remove background noise and to represent the entire dynamics of the LRM process at a given moment. By increasing the image threshold level, the sensitivity and informativity of the TG image adjusts to specific components of the image, which can be ideally approximated by a circular or ellipsoidal Gaussian distribution; for example, the 200-bit level will represent exclusively the highest amplitudes of within the TG image intensity.

Based on previous research [6, 9, 10, 17], a first order TF was chosen for identification of the $TF_{\text{LRM}}(s)$ and its parameters by calculating a Laplace transformation between $W(s)$ and $A_{\text{level}}(s)$ or $V_{\text{level}}(s)$ initially recorded in the time domain. Identification procedure was performed using the MATLAB signal processing toolbox taking synchronized in time $W(t)$, $A_{\text{level}}(t)$, $V_{\text{level}}(t)$ data inputs. After parameters of the $TF_{\text{LRM}}(s)$ (K and τ) being identified, experimentally obtained and calculated after identification $SRF_{\text{LRM}}(t)$ functions were compared and analyzed using a fit coefficient (%) and mean square error (bits) parameters with respect to each power and exposure time values.

4 Results and Discussion

The results of the identified TFs are shown in Tables 1 and 2 for calculated area and volume parameters at the 10-bit level. Using a model of TF with identified parameters, modelled SRFs were calculated and plotted against experimental SRFs in Fig. 3 for all exposure times of 50, 100, 150, and 200 μs. The fit and mean square error (MSE) between each pair of SRFs included in Tables 1 and 2 for estimation of the identification quality. The fits between pairs range from 73.75 to 96.36% confirming reliability of the identification procedure and the obtained data. Using pairs of SRFs at different bit levels and laser powers, the effect of the exposure times on reliable and informative demonstration of LRM dynamics was analyzed.

The laser power and exposure time both have a clear proportional relationship with the identified gain, while the rate of increase is independent of each other and varies for area-at-level and volume-above-level. The identified gains for each area-at-level TFs ranged from 1.59 to 29.71 pixel/W, where changes in bit level can dramatically

Table 1 Summary of identified parameters for area-at-level at 10-bit

Power (W)	Exposure (μs)	Timestep (μs)	Gain pixel/W	τ (μs)	Delay (steps)	Fit (%)	MSE (bits)
7.5	50	156.8	4.25	847.7	5	78.94	11.61
	100	206.7	9.24	706.0	2	84.90	27.69
	150	222.7	13.93	541.8	4	87.01	46.36
	200	253.6	17.76	491.0	2	90.01	44.34
10	50	156.8	12.16	417.0	2	92.35	21.62
	100	206.7	19.56	427.2	3	94.78	25.93
	150	222.7	24.14	385.0	2	94.38	45.81
	200	253.5	27.62	325.8	2	95.64	36.18
12.5	50	156.7	15.72	121.5	3	94.17	32.80
	100	206.7	22.71	144.3	2	95.62	38.53
	150	222.7	26.97	178.1	2	95.59	55.13
	200	253.6	29.71	220.4	2	96.36	45.65

Table 2 Summary of identified parameters for volume-at-level at 10-bit

Power (W)	Exposure (μs)	Timestep (μs)	Gain pixel/W	τ (μs)	Delay (steps)	Fit (%)	MSE (bits)
7.5	50	156.8	69.40	1102	5	73.75	4,913
	100	206.7	202.97	770.9	3	79.19	25,850
	150	222.7	367.11	742.1	4	81.69	64,880
	200	253.6	529.05	439.5	3	84.50	96,000
10	50	156.8	324.92	331.1	3	89.32	30,240
	100	206.7	753.79	632.4	3	92.04	89,700
	150	222.7	1180.18	709.4	2	91.78	234,300
	200	253.5	1590.30	590.0	2	93.50	265,800
12.5	50	156.7	538.55	139.3	3	92.68	60,820
	100	206.7	1203.28	178.4	2	94.74	156,000
	150	222.7	1766.18	215.7	2	95.11	291,400
	200	253.6	2260.57	321.1	2	95.20	459,000

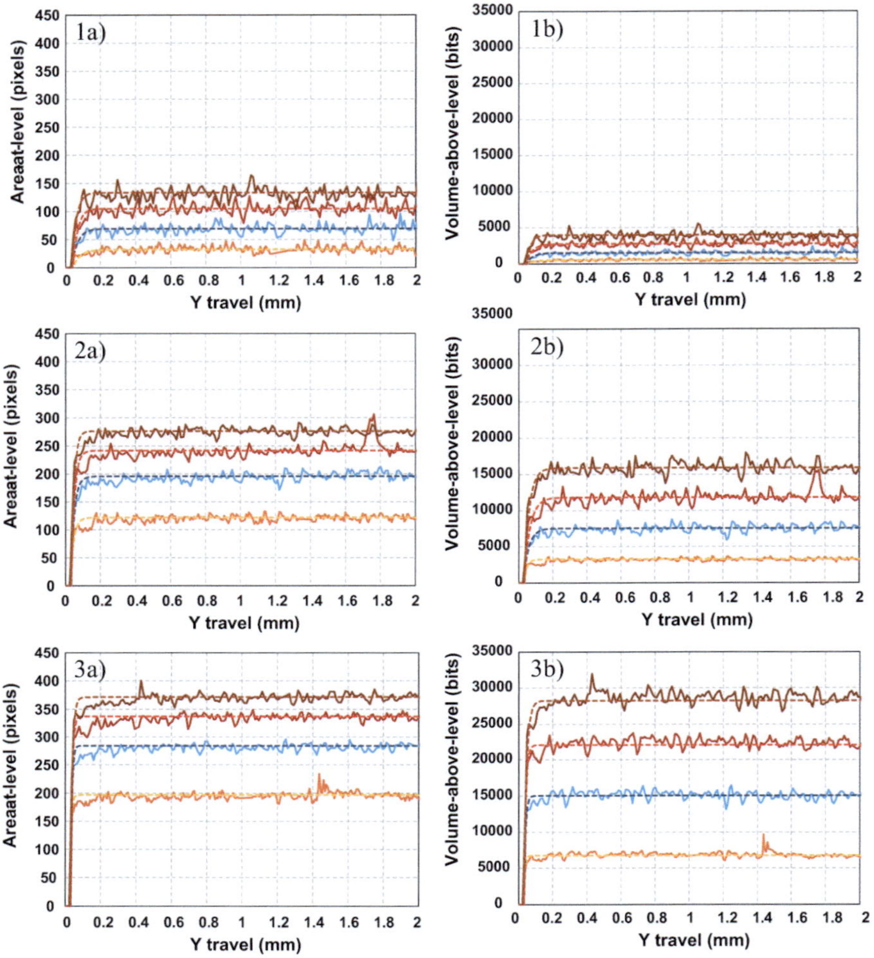

Fig. 3 Comparison of experimental and modeled SRFs for (1) 7.5 W, (2) 10 W, and (3) 12.5 W using **a** area-at-level and **b** volume-above-level parameters calculated at 10 bits, where the dashed lines represent the response of the identified SRFs and the solid lines represent the experimental SRFs obtained for 50 μs (yellow), 100 μs (blue), 150 μs (red), and 200 μs (brown)

swing the ranges. A bit level of 200 possesses a range of 1.59 to 3.54 pixel/W, in comparison to a bit level of 10 which has a range of 4.25 to 29.71 pixels/W. The identified gain for the respective volume-above-level TFs ranged from 69.4 to 2260.57 bits/W and demonstrated the same behavior in response to the changes in power, exposure, and bit level. The plotted SRF pairs demonstrate the same general behavior where the simulated response follows the recorded experimental data; this trend was observed for all identified TFs. The main difference between responses is the magnitude of the steady state response and the variance about that point. The plotted SRF clearly shows the increase in gain from increasing power and exposure.

SRF's capacity to achieve an appropriate fit was dependent on the ability of the bit level to retain enough useful information in the acquired image. More specifically, for 100-bit and 200-bit levels, the 7.5 W experiments yielded no identified TFs as the amplitudes of recorded TG images were below the bit level and significant distortions of data occur across the lower exposures. The comparison of the results between bit levels showed that there are small differences in the response time between the higher exposures, with the notable variances occurring close to the point where bit level starts to distort the response. The bit level had a larger impact on the system gain, showing an inverse relationship between this parameter and gain, where increasing the bit level removes more information from the signature calculation. Generally speaking, increasing the power and exposure time provided better fits and slightly faster response times. This is likely a consequence of the fact that higher magnitudes enable a clearer image of the process being tracked. For this reason, time constants are twice as low for the 12.5 W experiments than for the 7.5 W ones. Reducing exposure time enables faster recording of process dynamics but also decreases the amount of collected information. This highlights the primary challenge in choosing an exposure time: balancing the rate of information capture (for subsequent control) with the quality of data regarding the LRM process dynamics.

The 50 μs exposure time was the most versatile due to its ability to capture process dynamics, allowing for a large range of powers without overexposure and the fastest acquisition frequency. Conversely, the 50 μs response was not able to utilize the higher thresholds for image signatures leading to less informative representation of the process. By contrast, the 150 and 200 μs exposure times were able to capture images that completely filled the intensity range of the images and more complete and reliable information about temperature distribution and its gradients can be analyzed. This allowed different thresholds to be used in image signatures and provided a better fit to process dynamics. The performance of the 150 and 200 μs shows that they are specifically optimal for this chosen configuration of parameters, but any further power increases would result in oversaturation.

5 Summary and Conclusion

This research study describes the effect of the acquisition time while taking thermographic images by NIR camera on quality and informativity of the laser remelting process dynamics. It was shown that acquisition time constitutes the most critical component of the control cycle interval. The acquisition time also affects the observability and identifiability of LRM process dynamics represented by a first order linear system. Several experiments were performed to obtain thermographic (TG) step response functions (SRF) as a reaction on an instant change in applied laser power and recorded as a sequence of TG images acquired in both time and space domains. During experiments, three laser powers, 7.5, 10, and 12.5 W, and four acquisition times, 50, 100, 150, and 200 μs, were used. For each experiment, TG TF of the LRM process was identified using power step function as an input and TG SRF as an output of the continuous time system in linear approximation. Parameters of each TG TF were identified and SRF was simulated to be compared with experimentally obtained SRF. The quality of identification was estimated by a fit coefficient and a mean square error value.

The results obtained prompted the following conclusions:

- Power and exposure time both have a similar effect on the gain and response time of the modelled TFs, but act at different rates. Exposure time balances the acquisition rate and acquisition quality whereas the overall quality will improve with the power of the process.
- Increasing the threshold level applied to calculate area-at-level and volume-above-level of TG images leads to a decrease of the process gain, but as it increases, it reduces the process information and obscures the results of lower powers. The optimal exposure time is strongly influenced by other parameters to ensure maximum utilization of the intensity range, which is estimated by a volume-above-level parameter, without any losses. Additionally, higher exposure times are not well-suited for power levels exceeding 12.5 W that seemed to work best with a 50 μs exposure time. This enabled further captures of the natural intensity range and higher acquisitional frequencies. Introducing further changes to the process parameters would also increase the need for faster dynamics acquisition and greater intensity range. Therefore, 50 μs appeared to be the most suitable exposure time for processes that need to accommodate a wide range of powers. This exposure time also allows for further exploration into strategies that combine multiple images to increase the available data and maintain acquisition rate.

Future continuations of this study will attempt to expand the range of TG image parameters utilized and to investigate how combining images could increase the quality of information for the development of close-loop control systems of the LRM process.

References

1. Dobrzański LA, Dobrzańska-Danikiewicz AD (2019) Applications of laser processing of materials in surface engineering in the Industry 4.0 stage of the industrial revolution. Mater Perform Charact 8(6):1091–1129
2. Bordatchev EV, Hafiz AMK, Tutunea-Fatan OR (2014) Performance of laser polishing in finishing of metallic surfaces. Int J Adv Manuf Technol 73(1–4):35–52
3. Temmler A, Willenborg E, Wissenbach K (2012) Laser polishing. In: Laser applications in microelectronic and optoelectronic manufacturing (LAMOM) XVII, Proc. of the SPIE, pp 171–183
4. Pfefferkorn FE, Duffie NA, Li X, Vadali M, Ma C (2013) Improving surface finish in pulsed laser micro polishing using thermocapillary flow. CIRP Ann 62(1):203–206
5. Bordatchev E, Cvijanovic S, Wu H, Gorski A, Beyfuss D, Tutunea-Fatan RO (2023) Conceptualization and preliminary development of statistical digital twin and cyber-thermophysical system for advanced analysis, monitoring, and control of the laser remelting process. In: Proceedings of IEEE international conference on systems, man, and cybernetics, pp 3033–3039

6. Chow MTC, Hafiz AMK, Tutunea-Fatan OR, Knopf GK, Bordatchev EV (2010) Experimental statistical analysis of laser micropolishing process. In: Proceedings of international symposium on optomechatronic technologies, paper no. LPTI-1, 6 pp
7. Zou Z, Zhang K, Zhu Z, Liu T, Li J, Xiong Z, Li S, Liao W (2024) Multi-sensor monitoring of powder melting states via melt pool optical emission signals during laser-based powder bed fusion. Opt Laser Technol 169, paper 109880
8. Lane B, Moylan S, Whitenton EP, Ma L (2016) Thermographic measurements of the commercial laser powder bed fusion process at NIST. Rapid Prototyping J 22(5):778–787
9. Romer GRBE, Aarts J, Meijer RGKM (1999) Multivariable control of laser alloying of Ti6A14V. In: Proceedings of the ICALEO'99, Laser Institute of America, pp E43–E52
10. Shkoruta A (2021) Data-driven control of laser powder bed fusion. PhD thesis, Rensselaer Polytechnic Institute, 168 pp
11. Bordatchev EV, Nikumb SK (2006) Effect of focus position on informational properties of acoustic emission generated by laser material interactions. J Appl Surf Sci 253:1122–1129
12. Bordatchev EV, Cvijanovic S, Tutunea-Fatan OR (2019) Effect of initial surface topography during laser polishing process: statistical analysis. Procedia Manuf 34:269–274
13. Sun W, Zhang Z, Ren W, Mazumder J, Jin J (2022) In situ monitoring of optical emission spectra for microscopic pores in metal additive manufacturing. J Manuf Sci Eng 144(1), paper 011006
14. Chang Y-J, Tseng G-R, Ho C-C, Hsu J-C, Kuo C-L (2013) Detection of laser induced plasma with interdigital electrodes in laser material processing. Opt Lasers Eng 51(11):1199–1205
15. El Gamal A, Eltoukhy H (2005) CMOS image sensors. IEEE Circuits Devices Mag 21(3):6–20
16. Knaak C, Thombansen U, Abels P, Kröger M (2018) Machine learning as a comparative tool to determine the relevance of signal features in laser welding. Procedia CIRP 74:623–627
17. Meyer P, Bordatchev EV, Tutunea-Fatan OR (2024) Preliminary development of thermographic monitoring of the laser remelting process. In: Proceedings of CLEO'2024 conference paper AM1C.3, 2 pp

3D Mechanical Object Clustering for Shape Matching Using Multi-view Multi-slice CNN Architecture

Dipesh Shrestha, Arpan Man Sainju, and Abigail Kelly

Abstract The rapid growth of 3D object databases in mechanical engineering leads to the need for efficient shape-based clustering and retrieval methods. Accurate and fast clustering of 3D mechanical objects is vital for design reuse, similarity-based search, and optimizing manufacturing processes. This paper introduces a Multi-View Multi-Slice Convolutional Neural Network (MVMS-CNN) with semi-supervised learning for efficient 3D object clustering and shape matching. Our method automatically extracts complex features from multiple views and slices, enhancing accuracy and efficiency in shape retrieval for large databases. The MVMS-CNN captures both external and internal features of 3D objects. When combined with K-Means clustering, the MVMS-CNN demonstrates better purity and clustering accuracy than existing methods. The semisupervised learning approach also addresses the challenge of varied and unknown class numbers, making it well-suited for real-world scenarios involving large and diverse datasets. This combination allows for precise clustering, benefiting numerous applications in mechanical engineering and manufacturing.

Keywords 3D mechanical object clustering · Multi-view multi-slice · Convolutional neural network · Feature extraction

1 Introduction

The global Auto Parts Manufacturing Market is expected to expand at a CAGR (Compound Annual Growth Rate) of 6.8% between 2023 and 2030, reaching a value of around $1.1 trillion by 2030 [20]. The rapid expansion of 3D object databases in

D. Shrestha (✉) · A. M. Sainju · A. Kelly
Middle Tennessee State University, Murfreesboro, TN, USA
e-mail: ds2af@mtmail.mtsu edu

A. M. Sainju
e-mail: arpan.sainju@mtsu.edu

A. Kelly
e-mail: amk7r@mtmail.mtsu.edu

S. Y. Liang (ed.), *Manufacturing Technologies*, Lecture Notes in Mechanical Engineering, https://doi.org/10.1007/978-3-031-95455-9_11

mechanical engineering and manufacturing has led to the need for effective shape-based clustering and retrieval methods. Accurate and fast clustering of 3D mechanical objects is critical for many applications, such as cost estimation, design reuse, and similarity-based search. Traditional shape-matching methods frequently struggle with the complexity and diversity of mechanical parts, requiring more time and effort and being prone to human error.

Recent advancements in deep learning, particularly Convolutional Neural Networks (CNNs), have achieved great success in image recognition. Building on this, we propose a Multi-View Multi-Slice CNN (MVMS-CNN) with a semisupervised learning approach for efficient 3D object clustering and shape matching. Traditional methods often rely on hand-crafted features or simple geometric descriptors [4, 11], which may not capture the full complexity of 3D mechanical objects. Our approach aims to address the limitations of existing methods and provide an efficient solution for 3D mechanical object shape matching within vast and varied datasets.

A key motivation for this clustering method is efficient cost estimation in 3D object manufacturing, which is crucial during early product development and planning [16]. Objects with similar shapes generally have similar manufacturing costs due to shared production techniques and resources. Complex geometries associated with additive manufacturing (AM), such as 3D printing, often result in higher costs due to intricate processes, while simpler shapes in injection molding tend to be more cost-effective [9]. Clustering similar 3D objects allows manufacturers to quickly identify analogous parts and utilize existing cost data, resulting in faster and more accurate estimates. This research develops a deep learning-based approach to automatically find the best-matched 3D mechanical object from a diverse collection, using a semi-supervised learning approach. The key idea involves clustering the dataset of existing 3D objects and assigning clusters to new objects, enabling efficient shape matching.

However, automated matching of similar 3D objects presents unique challenges. First, identifying the best match from a large dataset is computationally intensive, especially with a continuous influx of new designs. Second, the diverse and unpredictable range of 3D object classes makes effective matching difficult. For instance, when dealing with a new 3D object, there are large amounts of data and varied classes to consider. Third, identifying complex shapes within 3D objects is challenging because some shapes are completely concealed and not observable externally. For example, internal cavities of a 3D object may be hidden, and recognizing this hidden information is crucial for efficient shape matching.

To address these challenges, our proposed approach utilizes semi-supervised learning with the MVMS-CNN network as a feature extractor. This method employs labeled data to train the feature extractor and incorporates unlabeled data for clustering, effectively managing diverse data classes. The MVMS network plays a crucial role in capturing subtle and hidden features from 3D objects, which is essential for handling the increasing complexity and variety of designs. *Contributions:* Our paper makes the following contributions:

1. We propose a novel framework using an MVMS-based feature extractor with K-Means clustering to identify the best-matched 3D object.
2. We extend the existing Tracepart dataset of 3D objects by a new class of data for our proposed semi-supervised machine learning approach.
3. We compare our approach with 3D CNN and 2D Projection baselines, achieving significant improvements in cluster purity and clustering accuracy.

The structure of the paper is as follows: Sect. 2 introduces the problem statement, Sect. 3 reviews related work, Sect. 4 describes our proposed approach, Sect. 5 provides experimental evaluation and discussion, and Sect. 6 presents conclusions and future research directions.

2 Problem Statement

Traditional shape matching methods rely on manual inspection and hand-crafted features, which are time-consuming, prone to error, and often miss complex or hidden shapes. Therefore, there is a clear need for a more efficient and adaptable method to automate shape matching accurately without extensive manual effort. *Problem Definition*: The problem involves clustering a collection of diverse 3D objects. The objective is to develop an efficient clustering algorithm that can categorize these 3D objects into distinct groups based on their similarity in shape. The input is a collection of diverse 3D objects, and the expected output is a set of clusters, each containing objects that are similar to one another. One of the main constraints is the potential size and diversity of the 3D object sets, which may exhibit varying levels of complexity and detail.

3 Related Work

3.1 3D Object Representation

3D objects can be represented in formats like Standard for the Exchange of Product data (STEP), mesh, point clouds, and voxel images. STEP is widely used in CAD (Computer-Aided Design) for precise 3D data exchange, using mathematical descriptions like Non-Uniform Rational B-Splines (NURBS) to represent complex geometries [3, 6]. Mesh formats like STereoLithography (STL) [5] and Object (OBJ) represent surface geometry but may lose geometric accuracy. Voxel images, used by Wu et al. [22] and Maturana and Scherer [14], simplify input but may suffer from high computational costs and quantization artifacts. We propose using binarized voxel images to represent 3D objects that offer a simple and interpretable representation while addressing limitations like high computational cost and quantization artifacts.

3.2 Feature Extraction for 3D Objects

3D CNNs have been used for feature extraction from voxelized 3D data [16, 23]. Graph-based approach [13] focuses on object classification but often overlooks geometric details. Bickel et al. [4] and Su et al. [19] used multi-view projections for shape retrieval, but these methods may fail to fully capture complex shapes. CurvaNet [10] uses geometric deep learning to extract curvature-based features from 3D mesh surfaces. Previous research [24, 25] demonstrated the effectiveness of multiview multi-slice techniques for 3D medical image analysis. In our work, we extend the use of MVMS-based approaches for feature extraction from binarized voxel images.

3.3 Clustering Methods

Clustering algorithms like K-means [7] are widely used for segmenting 3D data. Qi et al. [17, 18] applied PointNet architectures for clustering 3D point clouds. Liu et al. [12] and Wang and Lu [21] proposed hybrid methods like Point-Voxel CNN (PVCNN) and VoxSegNet, combining point-based and voxel-based approaches for efficient clustering and segmentation. These methods show the effectiveness of combining deep learning with clustering for 3D data segmentation.

4 Methodology

In the data-preparation phase, detailed in Sect. 5.1, 3D objects (samples in Fig. 1) are voxelized. The proposed methodology then comprises three phases:

I. Feature Extraction Phase We employ a MVMS-CNN and Principal Component Analysis (PCA) [1] for feature extraction. Figure 2 shows the architecture for feature extraction using the MVMS-CNN approach, while Fig. 3 depicts the overview of the proposed framework. First, the 3D voxel data is sliced at a certain 'slicing interval' from three views (X, Y, and Z). Each slice from each view is fed into a 2D CNN with a vector of fixed size (CNN Output Size) in its second to-last layer (the last layer is for classification into classes). The architecture of individual 2D CNN is depicted in Fig. 4. All these vectors are then concatenated to create a long feature vector for each voxel image. The dimensionality of the feature vector is subsequently reduced using the PCA algorithm, which helps in reducing dimensionality while retaining the most significant features.

II. Clustering Phase As shown in Fig. 3, the output from PCA is clustered using the K-Means clustering algorithm. It is important that the number of clusters is not set too high, as that could result in clusters containing only single elements. Such clusters would not be useful for our objective of matching 3D objects within the

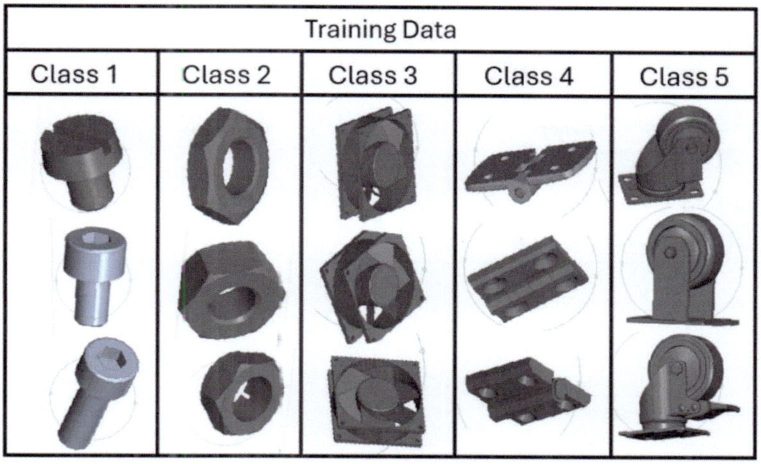

Fig. 1 Sample of the traceparts dataset

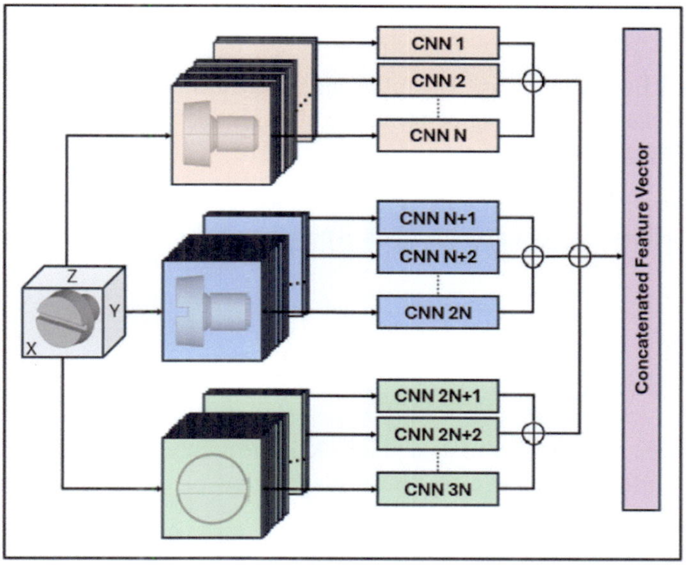

Fig. 2 MVMS CNN architecture

same cluster. The number of clusters shall be chosen to allow meaningful sub-clustering within each class. The resulting clusters will be used to determine the closest match to an unknown 3D object by selecting the nearest object within the same cluster.

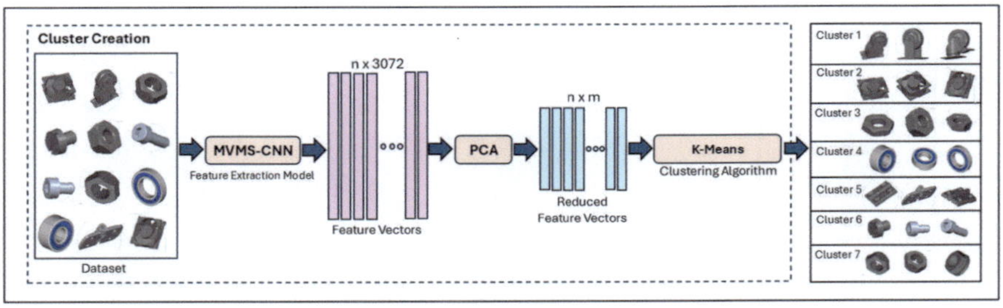

Fig. 3 Overview of proposed framework

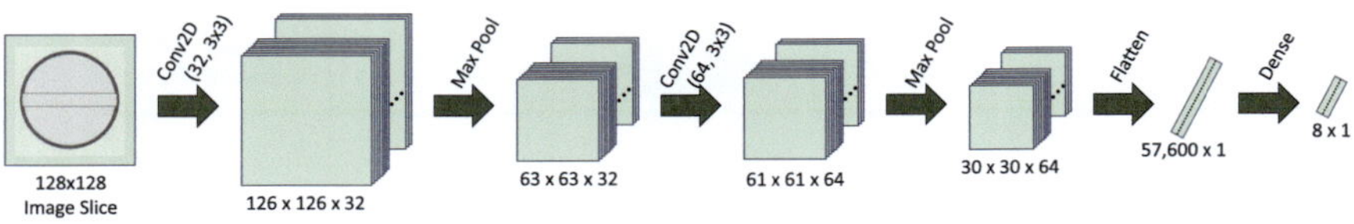

Fig. 4 Architecture of each CNN model with image slice of 128 × 128

III. Cluster Evaluation Phase The following cluster evaluation techniques are used to evaluate the clustering performance of K-Means Clustering methods on the Clustering Data.

Overall Purity: The overall purity is a measure of the extent to which each cluster contains data points from a single class. The formula for overall purity is given in Eq. 1.

$$\text{Overall Purity} = \frac{1}{N}\sum_{k=1}^{K}\max_{j} n_k^{(j)} \tag{1}$$

where N is the total number of data points, K is the number of clusters, and $n_k{}^{(j)}$ is the number of data points of class j in cluster k.

Clustering Accuracy: Clustering accuracy is a measure of how well the assigned clusters correspond to the ground-truth class labels. Equation 2 is the formula for clustering accuracy [8].

$$\text{Clustering Accuracy} = \frac{1}{N}\sum_{i=1}^{N}\delta(y_i, \text{map}(c_i)) \tag{2}$$

where N is total number of data points, y_i is true label of data point i, c_i is cluster label assigned to data point i, map(c_i) is the mapping function from clusters to true labels, based on the highest overlap in our case, and $\delta(y_i,\text{map}(c_i))$ is an indicator function that returns 1 if $y_i = \text{map}(c_i)$, and 0 otherwise.

5 Experimental Evaluation and Discussion

In this section, we present a comprehensive experimental evaluation of our proposed approach, including hyperparameter tuning and comparative analysis of MVMS-CNN with 3D CNN and 2D Projection feature extraction approaches. For comparison, we included a 3D CNN model and a 2D Projection model with similar number of layers. For the 3D CNN, we directly input the 3D voxel image, whereas for the 2D Projection model, we extracted 2D views from the x, y, and z axes and fed each view to a 2D CNN model.

5.1 Dataset and Experimental Setup

The primary dataset used for this research is the Traceparts Dataset [13], consisting of STEP files categorized into six classes, each with approximately 100 3D objects, which were converted to voxel images. A sample of this dataset is shown in Fig. 1. We extended this dataset with 57 images of the 'bearing' class that we collected. Different segments of the dataset are utilized for training the feature extraction model and clustering. The Traceparts dataset (denoted as $A + C$) and the new 'bearing' class dataset (denoted as $B + D$) are partitioned as follows: Training Data (A), used to train and test the MVMS-CNN model with an 80/20 split; Clustering Data ($A + B$), which combines samples from both datasets; and Testing Data ($C + D$), consisting of 20 images per class, used to evaluate clustering accuracy. The clustering data ($A + C$) is prepared by augmenting the training data (A) with a portion of new data from the 'bearing' class (C). The goal is to develop a feature extraction model using a subset of the dataset and then perform clustering on the augmented data, which includes previously unseen classes. As a result, the clustering data comprises seven classes.

First, each 3D object is converted into a binarized voxel image of size 128 × 128 × 128. For STEP file formats, the STEPToPoints tool [2] is used to transform the object into a point cloud, which is then processed into a voxel image using the pyntcloud Python library [15], preserving only geometric information. The clustering data is processed through the trained MVMS model, and the output from the second-last layer is fed into PCA for dimensionality reduction. The number of PCA components is selected so as to explain at least 70% of the variance, as determined by the scree plot. Choosing a higher percentage would require significantly more components. For our dataset, capturing 70% of the variance provides an effective balance between dimensionality reduction and retaining essential variability. The feature vectors obtained from PCA are clustered using the K-Means clustering algorithm. The number of clusters is chosen to allow meaningful sub-clustering within each class while avoiding over-segmentation. Specifically, we selected 20 clusters—approximately three times the number of classes. Selecting 20 clusters provides a balance, allowing for meaningful subclustering without over-segmenting the data.

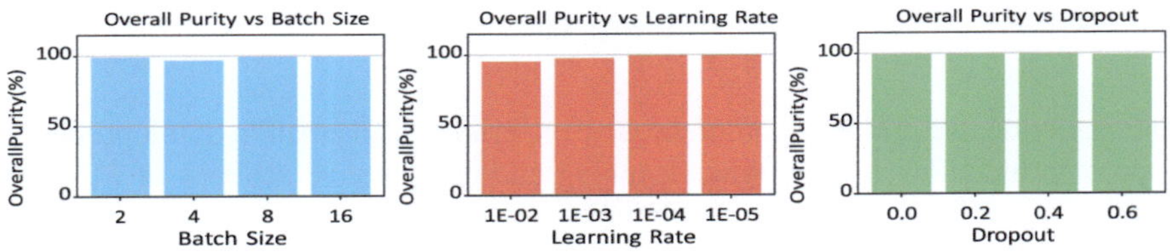

Fig. 5 Hyperparameter tuning for CNN training

5.2 Hyperparameter Tuning for CNN Training

We initiated our experimental phase by fine-tuning the hyperparameters for CNN Training, namely Dropout Ratio, Learning Rate, and Batch Size, for the MVMS-CNN with K-Means clustering methods. These experiments were conducted using 40 epochs, a slicing interval value of 1, and a cube size of $128 \times 128 \times 128$ for a CNN output vector size of 8×1. An ablation study was conducted with a default batch size of 8, learning rate of 0.001, and dropout rate of 0%, varying one parameter at a time while keeping the others fixed. Our focus was on achieving high overall purity, as this metric provides a clear indication of the homogeneity of the clusters and helps ensure that the data points within each cluster are highly similar to one another.

The summarized results of these experiments are presented in Fig. 5. We observe that batch size of 8 and 16 have highest overall purity of 99.6%. For computational cost effectiveness, batch size of 8 was designated as optimal choice. Further, model achieved highest purity (99.6%) at learning rate of 1E-4 and 1E-5. Lower learning rate of 1E-4 was selected as the optimal value. Also, the dropout ratio appeared to have minimal impact on overall purity, and thus a value of 0 was selected to maximize model simplicity.

5.3 Hyperparameter Tuning for MVMS-CNN Architecture

We systematically experimented with different input cube sizes, slicing intervals, and CNN output sizes for the MVMS-CNN approach, aiming to optimize these hyper-parameters. We used overall purity and clustering accuracy as evaluation metrics to guide our optimization process.

Cube Size Fig. 6a, d shows that increasing cube size generally led to improvements in overall purity and clustering accuracy, with a cube size of 128 yielding the highest overall purity of 99.6% and clustering accuracy of 98.5%. This suggests that larger cube sizes enable the model to capture more detailed structural information, crucial for accurate clustering. We did not explore cube sizes beyond 128 as it was the size of voxel image used. Also, larger cube sizes (e.g., 256) would be computationally prohibitive. Since the highest performance was achieved with a cube size of 128, this value was selected for further analysis.

CNN Output Size (CNN_Op_Size): According to Fig. 6b, e, the overall purity and clustering accuracy are less sensitive to variations in CNN output sizes. This low sensitivity can be attributed to applying PCA after CNN, effectively engineering the most important features from these outputs. Generally, larger output sizes offer slightly better performance, but at a higher computational cost due to increased space complexity. Balancing performance with computational efficiency, a CNN output size of 8 was selected for further analysis.

Slicing Interval: As depicted in Fig. 6c, f, overall purity for different slicing intervals indicates that a slicing interval of 1, 2 and 4 achieved the highest purity of 99.6%. Also, the highest clustering accuracy of 98.5% was observed for a slicing interval of 1, leading us to select this slicing interval for further analysis. Generally, overall purity and clustering accuracy seem to be less sensitive for smaller slicing interval values but tend to decrease with increasing slicing interval. This is because a lower slicing interval value ensures analysis of a greater number of slices of data, retaining more information. Conversely, increasing the slicing interval skips slices, resulting in information loss and reduced performance.

5.4 *Comparative Analysis*

Finally, we compared the results of the MVMS-CNN approach with the 3D CNN and 2D Projection approaches. The comparison focused on overall clustering purity and clustering accuracy. The experiment used K-Means as the clustering algorithm, with 40 epochs, a learning rate of 1×10^{-4}, a slicing interval of 1, a CNN output size of 8, and a cube size of 128. To ensure fairness and consistency, the same set of hyperparameters was used for all models. This approach ensures that the performance improvements observed are due to inherent architectural differences rather than selective tuning.

Table 1 shows that the MVMS-CNN approach achieved an overall purity of 99.6% and clustering accuracy of 98.5%. These values are higher than those achieved by the 3D CNN and 2D Projection methods. The superior performance of the MVMS-CNN approach compared to the 3D CNN and 2D Projection methods can be attributed to its ability to integrate information from multiple views and slices, leading to a more comprehensive representation of 3D objects. This enhanced feature extraction capability makes the MVMS-CNN particularly effective for detailed shape analysis tasks. Thus, resulting in higher clustering purity and clustering accuracy.

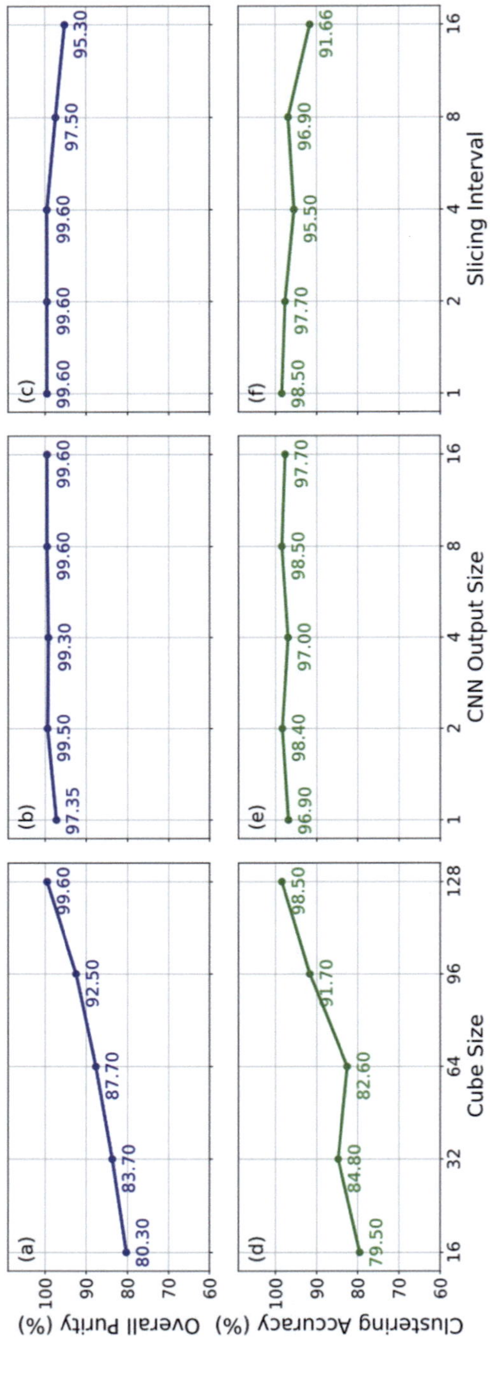

Fig. 6 Hyperparameter tuning for MVMS-CNN architecture

Table 1 Comparison of MVMS, 3DCNN, and 2D projection

Method	Purity (%)	Clustering accuracy (%)
MVMS	99.6	98.5
3D CNN	98.0	93.9
2D projection	94.1	92.4

6 Conclusion and Future Works

In conclusion, our experimentation demonstrates the effectiveness of the MVMS-CNN approach with K-Means clustering for shape matching. Systematic hyperparameter tuning optimized the model for high clustering purity and clustering accuracy. MVMS-CNN's integration of multiple views and slices provided superior overall purity and clustering accuracy. It outperformed 3D CNN and 2D Projection methods, demonstrating its ability to extract more distinctive and informative features from 3D mechanical objects. For future research, we will explore a hybrid approach that integrates MVMS-CNN with contrastive learning. We will also investigate the incorporation of Vision Transformers (ViT) for feature extraction and representation of 3D objects.

References

1. Abdi H, Williams LJ (2010) Principal component analysis. Wiley Interdiscipl Rev Comput Stat 2(4):433–459
2. Aleutgeb A (2024) Steptopoints: the program steptopoints converts solids contained in step files into point clouds by regular sampling. https://github.com/aleutgeb/STEPToPoints
3. Baumgart BG (1975) A polyhedron representation for computer vision. In: Proceedings of the national computer conference and exposition, pp 589–596
4. Bickel S, Schleich B, Wartzack S (2023) A novel shape retrieval method for 3d mechanical components based on object projection, pre-trained deep learning models and autoencoder. Comput Aided Des 154:103417
5. Crump SS (1992) Apparatus and method for creating three-dimensional objects, June 9 1992, US Patent 5,121,329
6. De Lazzari E, van den Boom SJ, Zhang J, van Keulen F, Aragón AM (2021) Acritical view on the use of non-uniform rational b-splines to improve geometry representation in enriched finite element methods. Int J Numer Meth Eng 122(5):1195–1216
7. Ding C, He X (2004) K-means clustering via principal component analysis. In: Proceedings of the twenty-first international conference on Machine learning, p 29
8. Fränti P, Sieranoja S (2024) Clustering accuracy. Appl Comput Intell 4(1):24–44
9. Gibson I, Rosen D, Stucker B (2015) Additive manufacturing technologies: 3D printing, rapid prototyping, and direct digital manufacturing, 2nd ed, chap. 12. Springer, New York, NY, pp 387–393. https://doi.org/10.1007/978-1-4939-2113-3
10. He W, Jiang Z, Zhang C, Sainju AM (2020) Curvanet: geometric deep learning based on directional curvature for 3D shape analysis. In: Proceedings of the 26th ACM SIGKDD international conference on knowledge discovery and data mining, pp 2214–2224
11. Jayanti S, Kalyanaraman Y, Ramani K (2009) Shape-based clustering for 3D cad objects: a comparative study of effectiveness. Comput Aided Des 41(12):999–1007

12. Liu Z, Tang H, Lin Y, Han S (2019) Point-voxel CNN for efficient 3D deep learning. In: Conference on neural information processing systems (NeurIPS)
13. Mandelli L, Berretti S (2022) Cad 3D model classification by graph neural networks: a new approach based on step format (10, 2022). http://arxiv.org/abs/2210.16815
14. Maturana D, Scherer S (2015) Voxnet: a 3D convolutional neural network for realtime object recognition. In: 2015 IEEE/RSJ international conference on intelligent robots and systems (IROS). IEEE, pp 922–928
15. Nikolov I (2022) How to voxelize meshes and point clouds in python. Towards Data Science
16. Ning F, Shi Y, Cai M, Xu W, Zhang X (2020) Manufacturing cost estimation based on a deep-learning method. J Manuf Syst 54:186–195
17. Qi CR, Su H, Mo K, Guibas LJ (2017) Pointnet: deep learning on point sets for 3D classification and segmentation. In: Proceedings of the IEEE conference on computer vision and pattern recognition, pp 652–660
18. Qi CR, Yi L, Su H, Guibas LJ (2017) Pointnet++: deep hierarchical feature learning on point sets in a metric space
19. Su H, Maji S, Kalogerakis E, Learned-Miller E (2015) Multi-view convolutional neural networks for 3d shape recognition. In: Proceedings of the IEEE international conference on computer vision, pp 945–953
20. Vantage Market Research: Global auto parts market size & share to surpass $1103.4 billion by 2030: Vantage Market Research. GlobeNewswire (November 2023)
21. Wang Z, Lu F (2019) Voxsegnet: volumetric CNNs for semantic part segmentation of 3D shapes. IEEE Trans Visual Comput Graphics 26(9):2857–2866
22. Wu Z, Song S, Khosla A, Yu F, Zhang L, Tang X, Xiao J (2014) 3dshapenets: a deep representation for volumetric shapes (6, 2014). http://arxiv.org/abs/1406.5670
23. Yoo S, Kang N (2021) Explainable artificial intelligence for manufacturing cost estimation and machining feature visualization. Expert Syst Appl 183
24. Zeng X, Chen S, Xie Y, Liao T (2023) 3v3d: Three-view contextual cross-slice difference three-dimensional medical image segmentation adversarial network. ACM Trans Multimed Comput Commun Appl 19(6):1–28
25. Zhang Y, Peng S, Xue Z, Zhao G, Li Q, Zhu Z, Gao Y, Kong L (2023) AMSF: attention-based multi-view slice fusion for early diagnosis of Alzheimer's disease. PeerJ Comput Sci 9:e1706

Target Detection and Tracking Method in Industrial Environment Based on TA-YOLOv8s

Wenkai Bao, Hui Ma, Baisen Liu, and Jiahui Shi

Abstract In modern industrial environments, target detection and target tracking play an important role. When dealing with complex industrial scenes, the current popular target tracking algorithms have the problem of insufficient tracking accuracy for objects with the same features and low robustness under occlusion. This paper introduces TA-You Only Look Once version 8 small (TA-YOLOv8s). This model introduces the Triplet Attention module in the YOLOv8s framework, captures cross-dimensional interactions through three branch structures, and contains channel information and spatial information at the same time. It overcomes the problems of channel and space separation calculation and failure to consider cross-dimensional interactions and dimensionality reduction in common attention methods, and can better express network features. At the same time, a backtracking algorithm (Recall Track) is proposed. To solve the problem of tracking loss caused by occlusion. Experimental results show that our method achieves an accuracy of 90.5% for target tracking when the occlusion ratio is 36%, surpassing the existing technology.

Keywords Occluded object tracking · YOLOv8 · Attention mechanism · Object detection · Deep learning

---

W. Bao · H. Ma (✉)
School of Electronic Engineering, Heilongjiang University, Harbin, China
e-mail: 2011043@hlju.edu.cn

W. Bao
e-mail: baowenkai@s.hlju.edu.cn

B. Liu · J. Shi
School of Electronic and Information Engineering, Heilongjiang Institute of Technology, Harbin, China

S. Y. Liang (ed.), *Manufacturing Technologies*, Lecture Notes in Mechanical Engineering, https://doi.org/10.1007/978-3-031-95455-9_12

1 Introduction

In the industrial field, the core content of tracking tasks includes real-time identification and positioning of multiple targets, handling occlusion, and distinguishing similar targets. Unlike traditional detection tasks, detection tasks refer to identifying and locating all targets of interest in a single frame of image or video frame, and outputting the bounding box and category label of each target. During the processing, one frame of image is processed each time, and the target is independently identified and located. The motion information of the target is not involved, and only the current frame is analyzed. The target tracking task is to continuously track the target in the video sequence, output the position of each target in each frame, and maintain the consistency of the target identity. In the actual industrial environment, the target tracking task needs to pay special attention to:

Handling occlusion problems, after a short occlusion, the algorithm must be able to quickly resume target tracking and maintain the consistency of the target identity [1]. Target similarity processing. When multiple targets with similar appearance appear at the same time, the target detection algorithm can detect them in a timely and accurate manner. The tracking algorithm needs to accurately distinguish and maintain the identity of each target. Moreover, due to the different tracking tasks in different industrial application environments, the tracking algorithm also needs to make corresponding adjustments as the task target changes. For long-term occlusion and the situation where the tracked target is extremely similar or completely similar, there is currently no particularly effective method to deal with it [2].

The main contributions of this paper are:

First, we use the Triplet Attention [3] module to optimize the target detection network model of YOLOv8. The three-branch structure captures cross-dimensional interactions, including channel information and spatial information at the same time, overcoming the problems of channel and space separation calculation and failure to consider cross-dimensional interactions and dimensionality reduction in common attention methods. Compared with some traditional attention mechanisms, it can better express network features and extract deeper features. It effectively solves the problem of identifying similar objects during tracking.

Secondly, we also proposed a backtracking method (Recall Track) for post-processing the output results of the tracking algorithm and applied this method to TA-YOLOv8. The method includes four steps: coding, binding, backtracking, and output. Among them, "coding" adds a unique identity information (QR Code) to each detected target, which is the core of judging the tracking result. "Binding" is to associate the coding content with the tracking ID. "Backtracking" is to trace back the previous corresponding ID and QR Code through the timeline when tracking is lost or severely occluded, and finally output the result.

2 Related Work

2.1 A Subsection Sample

The core of the target tracking task is to maintain the consistency of the target identity and accurately estimate the target position by continuously monitoring and analyzing the targets in the video sequence [4]. SORT is a lightweight, real-time multi-target tracking algorithm that mainly relies on Kalman filtering and Hungarian algorithm for target tracking and data association [5]. However, SORT only relies on the motion information of the target, which makes it easy to fail in complex scenes, especially when the targets are similar in appearance or occluded. To solve these problems, DeepSORT introduces deep learning technology, which greatly improves the robustness of the algorithm by combining visual features for tracking, and enhances the tracking effect in the case of similar target appearance and occlusion. Due to the introduction of deep learning models, the computational complexity and resource consumption are higher than traditional SORT. Despite the improvements, tracking failures may still occur in the case of long-term or frequent occlusion.

Lin et al. improved a real-time MOT method based on ByteTrack's fast appearance features, added new appearance features on the basis of ByteTrack [6], and re-matched some object boxes with low confidence to improve the matching progress, thereby reducing the occlusion problem [4]. Although this method does have a significant improvement effect on some situations with short occlusion time, the targets are all tasks with different features, and there is no effective improvement when dealing with similar targets or targets with the same appearance. In order to solve the problem of dealing with target similarity, Yang et al. proposed a new cascade buffer IoU (C-BIoU) tracker for tracking targets with indistinguishable appearance and high similarity [7]. However, the robustness of the tracker is low. In the face of complex environments in industrial environments, various noise interferences will lead to the stability of its work, thereby affecting the efficiency of the entire task.

3 Methodology

We use the powerful object detection network of YOLOv8 to detect objects of interest from the input image or video frame and return the bounding box, category label and confidence score of each object [8]. Then we use tracking algorithms, such as ByteTrack, BotSort [9], DeepSort [10], which are powerful multi-object tracking algorithms, to track the detected objects in the video sequence so that each object maintains a unique identity in consecutive frames. The overall framework of TA-YOLOv8 is shown in Fig. 1.

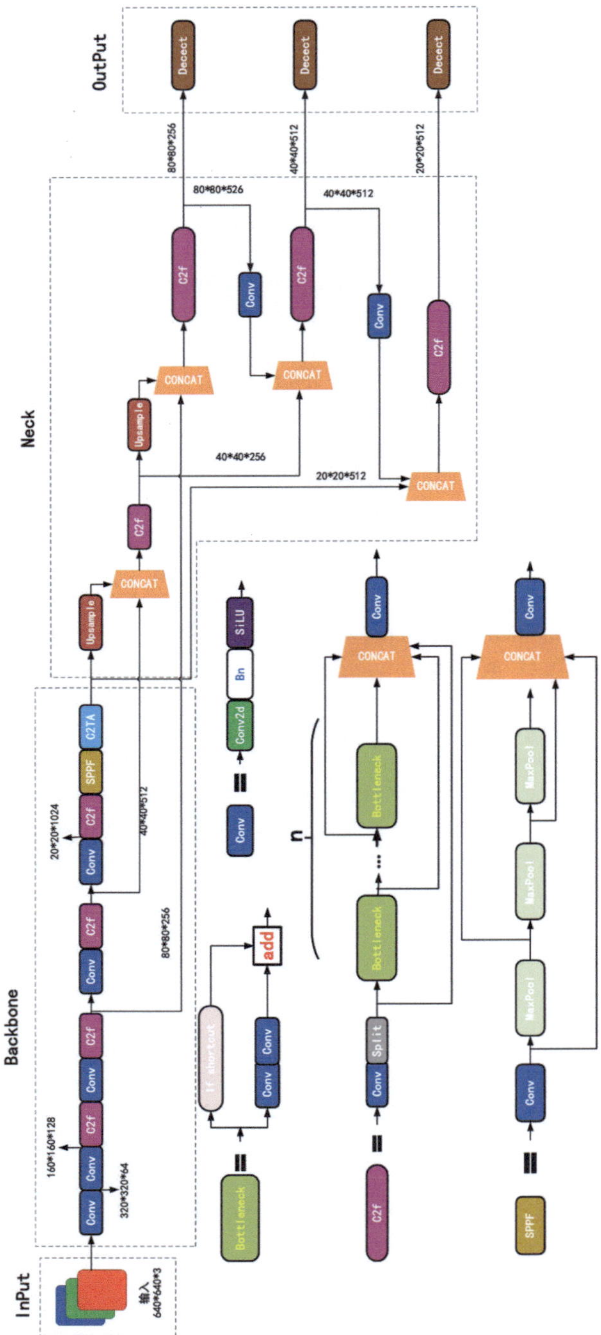

Fig. 1 TA-YOLOv8 overall framework

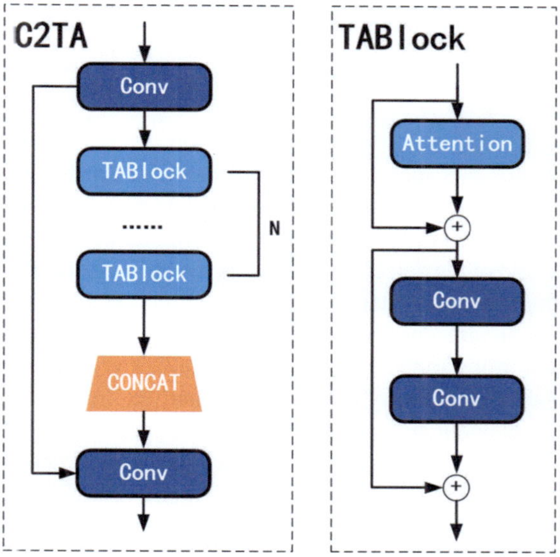

Fig. 2 C2TA structure diagram

3.1 *Triplet Attention Mechanism*

In this study, we use Triplet Attention, a new method for calculating attention weights by capturing cross-dimensional interactions using a three-branch structure. For the input tensor, Triplet Attention performs a rotation operation and then uses a residual transformation to establish dependencies between dimensions and encode inter-channel and spatial information with negligible computational overhead. This can better express the characteristics of the network. The architecture of C2TA is shown in Fig. 2.

3.2 *Recall Track Module*

In the tracking algorithm part, we proposed an adaptive threshold adjustment mechanism based on ByteTrack to optimize the performance of the algorithm in different scenarios. This mechanism dynamically adjusts the threshold according to the changes in the detector output results and the scene characteristics, thereby improving the accuracy and robustness of target tracking. Specifically, by analyzing the confidence distribution of the detector output and the relevant characteristics of the scene changes. At the same time, we designed a tracking backtracking module, which mainly includes: threshold adaptation, matching correlation, and ReID binding. Among them, threshold adaptation can adjust the threshold in real time to adapt to the diversity and complexity of the environment. This method not only enhances the system's adaptability to different environmental conditions but also improves the reliability of target recognition. The adaptive threshold formula is as follows:

$$\alpha = \frac{L_t}{W_v} \times \frac{F_r}{V_t} \times P_t \tag{1}$$

where α is the confidence threshold, which is the confidence parameter of Byte-Track, L_t is the pixel length of the detection target, W_v is the acquisition width of the video stream, F_r is the acquisition frame rate of the video stream, V_t is the movement speed of the detection target, and the last P_t is the motion trajectory parameter, which depends on the target movement direction and is generally 0.5–0.7. The detection results are divided into high confidence and low confidence parts for processing according to the confidence threshold. The high confidence detection results establish a preliminary tracking path, and the low confidence detection results need to supplement and optimize the tracking path.

Association matching is the core used to determine the degree of association between the lost ID and the newly generated ID. During the tracking task, if there is no loss of the tracking ID, or if the algorithm finds the tracking object again after a short disappearance, there is no need to associate the matching degree. When the ID is lost and the tracking trajectory cannot be retrieved through the tracking algorithm, it is necessary to start judging the last moment before the tracking trajectory is lost and record the location information of the target object. As the tracking task continues, a new tracking track, that is, a new ID, will be generated. At this time, starting from the moment the new track and new ID appear, the location information of the target object is recorded again. The two positions are calculated to traverse the entire cache area to find the information with the highest correlation for determination.

This equation represents the calculation of matching association:

$$D = \sqrt{(x - p_x)^2 + (y - p_y)^2} \tag{2}$$

where D is the output judgment distance, x and y represent the center position of the target bounding box when generating a new ID and a new trajectory, respectively, and p_x and p_y are the center positions of the bounding box of the target object at the time of loss

$$\eta = \frac{|D|}{d_{\text{set}}} \tag{3}$$

where d_{set} is the distance parameter to be compared with the distance to be determined $|D|$. The output of this equation will represent the confidence of the association.

Every time an occlusion occurs, the result will be saved in the buffer. According to the setting of the buffer memory space size, a certain number of results will be stored. When the buffer is full, the subsequent results will replace the results recorded at the beginning.

$$D_i = \sqrt{(x_i - p_x)^2 + (y_i - p_y)^2} \tag{4}$$

where D_i in i represents matching with the i-th result in the cache.

$$\underset{0 \le i \le u}{\eta_u} = \frac{|D_i|}{d_{\text{set}}} \tag{5}$$

After that, η_u represents all the results of the calculation. The best result is selected from η_u according to the confidence level and finally output. ReID rebinding is to store the image results processed within a period of time. We have opened up a space in the computer's memory space, the size of which depends on the number of results that need to be stored. The stored results mainly include various types of data after YOLOv8 detection, as well as the tracks output by the tracking algorithm and the identity ID of each track, as well as the QR Code bound to each track. That is to say, each output result in the cache contains the above results.

During the tracking task, when a slight occlusion occurs, the tracking algorithm is recoverable for the tracking trajectory. The results caused by the occlusion at this time are temporarily lost and will not be recorded. After an irreversible occlusion occurs, the tracking algorithm will regenerate a new trajectory. At this time, there is no QR Code on the newly generated trajectory. By finding the tracking results recorded before the loss in the Recall Track Buffer, the QR Code is re-extracted and assigned to the newly generated trajectory again. Finally, the processing results after backtracking are output. The execution process of Recall Track is shown in Fig. 3.

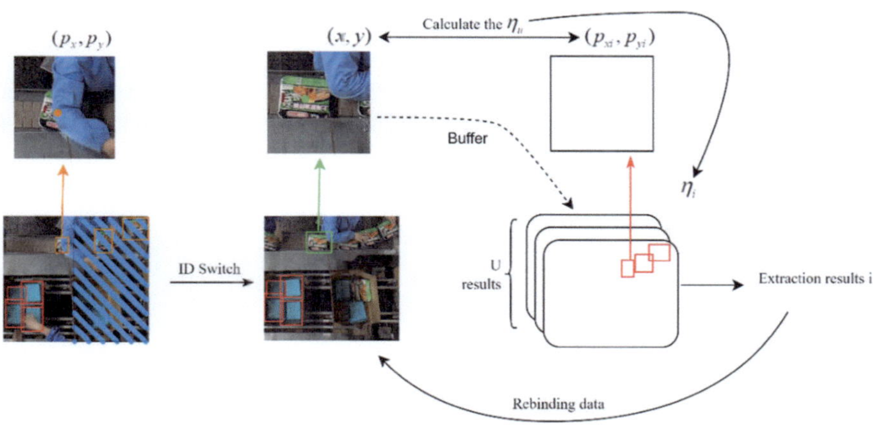

Fig. 3 Recall track flow chart

4 Methodology

4.1 Dataset

The dataset used in this experiment was collected and annotated by our team. Compared to widely used public datasets, this dataset has unique characteristics.

It was collected in real and complex industrial environments, containing actual production data under various operational conditions. The complexity and diversity of these scenarios far exceed the coverage of many public datasets. The data collection plan was designed based on specific industrial application needs, ensuring that it reflects common real-world challenges such as lighting variations, equipment vibrations, background complexity, and object occlusions during the production process. This dataset enables our research to better simulate and address object detection and tracking issues in industrial settings. The dataset consists of 40,870 images, covering products from seven different industrial production environments. All products within a category have the same size, with dimensions of 180 mm in length, 160 mm in width, and 65 mm in height. Due to the perspective of data collection, height information is not used. All images were captured from a perspective above the product conveyor belt, with a resolution of 2560×1440 pixels. During the annotation process, we strictly adhered to industrial standards and practical application requirements to ensure the accuracy of each annotation. All annotated data underwent multiple rounds of manual review and verification to ensure high data quality. The annotations include details such as object position, category, occlusion, and interactions with the background, ensuring that the algorithm is trained and validated comprehensively.

Due to the involvement of proprietary information from the industrial enterprise during data collection, we are unable to share the dataset publicly. However, researchers with legitimate needs can contact the corresponding author to request access. Figure 4 shows some sample images from the dataset, with annotated objects. Each sample represents a typical scenario in the actual production environment, including various lighting conditions, background complexities, and occlusions.

4.2 Comparison to Prior Work

We used YOLOv5n, YOLOv7, and TA-YOLOv8s as target detectors. The parameter performance of the model is shown in Table 1.

The performance of each model developed is evaluated by using precision and recall rate as shown in (6) and (7). TP, FP, TN, and FN stand for true positive, false positive, true negative, and false negative.

Fig. 4 Partial dataset

Table 1 Training performance of different YOLO detectors on the dataset

Model	P (%)	R (%)	mAP@0.5 (%)	mAP@0.5:0.95 (%)
YOLOv5n	95.1	92	96.15	83.12
YOLOv7	*96.2*	*97*	**98.83**	*83.85*
TA-YOLOv8s	**97.4**	**99**	*98.32*	**92.78**

We also selected DeepOCSORT, Improved Association Tracker (ImprAssocTrack), OCSORT, StrongSORT, BoT-SORT, and ByteTrack for comparative experiments. First, we used the three detectors and ByteTrack tracker for tracking without adding identity IDs to the tracked targets.

As shown in Fig. 5 (a) is the detection result of the YOLOv5n target detector, (b) is the detection result of the YOLOv7 target detector, and (c) is the detection result of the TA-YOLOv8s target detector.

From Fig. 5, we can see that both YOLOv5n and YOLOv7 have missing detections and detection errors.

As shown in Table 2, the three target detectors YOLOv5n, YOLOv7, and TA-YOLOv8s use DeepOCSORT, Improved Association Tracker (ImprAssocTrack), OCSORT [11], StrongSORT [12], BoT-SORT, and ByteTrack to track and detect the same video stream. Due to the detection performance of the detector, when using YOLOv5n [13] as the detector, although it is the fastest in processing speed,

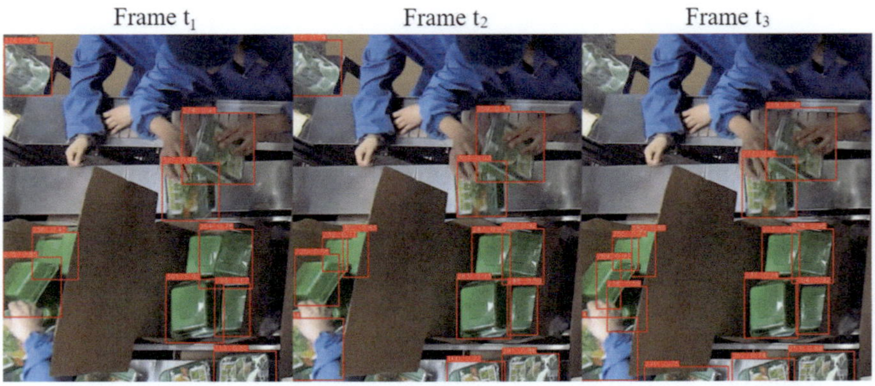

(a) Detector YOLOv5n (b) Detector YOLOv7 (c) Detector TA-YOLOv8s

Fig. 5 Tracking results of different detectors

its detection performance is undoubtedly the worst among the three compared to YOLOv7 [14] and TA-YOLOv8s. From the accuracy of the final tracking results, the improvement in accuracy is more obvious as the detection performance of the detector improves.

Table 2 Target detector and tracking algorithm statistics

Detection algorithm	Module			Result	
	Occlusion rate (%)	Tracker	ReID weights	FPS	Accuracy (%)
YOLOv5n ReCall track	36	DeepOCSORT	osnet_x0_25_msmt17	15	65.57
	36	ImprAssocTrack	osnet_x0_25_msmt17	18	62.07
	36	OCSORT	–	*23*	65.55
	36	StrongSORT	osnet_x0_25_msmt17	15	65.84
	36	BoT-SORT	osnet_x0_25_msmt17	17	**80.36**
	36	ByteTrack	–	**30**	*71.80*
YOLOv7 ReCall track	36	DeepOCSORT	osnet_x0_25_msmt17	13	77.59
	36	ImprAssocTrack	osnet_x0_25_msmt17	16	**89.08**
	36	OCSORT	–	*26*	78.57
	36	StrongSORT	osnet_x0_25_msmt17	14	*84.43*
	36	BoT-SORT	osnet_x0_25_msmt17	18	82.20
	36	ByteTrack	–	**27**	75.40
TA-YOLOv8s ReCall track	36	DeepOCSORT	osnet_x0_25_msmt17	10	84.17
	36	ImprAssocTrack	osnet_x0_25_msmt17	13	*86.89*
	36	OCSORT	–	*18*	83.76
	36	StrongSORT	osnet_x0_25_msmt17	12	85.72
	36	BoT-SORT	osnet_x0_25_msmt17	13	85.95
	36	ByteTrack	–	**20**	**90.50**

5 Conclusion

In this study, we aim to solve the problems of continuous occlusion in complex and dynamic environments, tracking interruption caused by display and backlight changes, and efficient identification and tracking of similar targets. We first introduced the Triplet Attention module into the backbone network of YOLOv8 and verified the improvement of the overall task by this module. Secondly, we proposed a backtracking algorithm called Recall Track to improve the efficiency and robustness of target tracking in complex and dynamic environments. For products with similar or identical features, we introduced a unique identity code (QR Code) for each tracked target in Recall Track to distinguish each similar target. For the problem of tracking loss caused by continuous occlusion and environmental reasons, we adopted a new correlation matching mechanism to retrieve the lost identity information to ensure the uniqueness of the detected target identity. In addition, our method has achieved a better balance between detection progress and speed among various detectors, which is very suitable for multi-target tracking applications in industrial environments.

Acknowledgements This work is partially supported by the Heilongjiang Provincial Natural Science Foundation (LH2022F047).

References

1. Cheng C-C, Qiu M-X, Chiang C-K, Lai S-H (2023) Rest: a reconfigurable spatial-temporal graph model for multi-camera multi-object tracking. In: Proceedings of the IEEE/CVF international conference on computer vision, pp 10051–10060
2. Shang Q, Zhang J, Yan G, Hong L, Zhang R, Li W, Xia H (2023) Target tracking algorithm based on occlusion prediction. Displays 79:102481
3. Misra D, Nalamada T, Arasanipalai AU et al (2021) Rotate to attend: convolutional triplet attention module. In: Proceedings of the IEEE/CVF winter conference on applications of computer vision, pp 3139–3148
4. Yang F, Odashima S, Masui S, Jiang S (2023) Hard to track objects with irregular motions and similar appearances? Make it easier by buffering the matching space. In: Proceedings of the IEEE/CVF winter conference on applications of computer vision, pp 4799–4808
5. Su S, Han S, Li Y, Zhang Z, Feng C, Ding C, Miao F (2024) Collaborative multi-object tracking with conformal uncertainty propagation. IEEE Robot Autom Lett
6. Zhang Y, Sun P, Jiang Y, Yu D, Weng F, Yuan Z, Luo P, Liu W, Wang X (2022) Bytetrack: multi-object tracking by associating every detection box. In: Avidan S, Brostow G, Cissé M, Farinella GM, Hassner T (eds) Computer vision–ECCV 2022, Springer Nature Switzerland, Cham, pp 1–21
7. Sun P, Cao J, Jiang Y, Yuan Z, Bai S, Kitani K, Luo P (2022) Dancetrack: multi-object tracking in uniform appearance and diverse motion. In: Proceedings of the IEEE/CVF conference on computer vision and pattern recognition, pp 20993–21002
8. Jocher G, Chaurasia A, Qiu J (2023) YOLO by ultralytics. Ultralytics
9. Aharon N, Orfaig R, Bobrovsky BZ (2022) BoT-SORT: robust associations multi-pedestrian tracking. arXiv:2206.14651
10. Maggiolino G, Ahmad A, Cao J, Kitani K (2023) Deep OC-sort: multi pedestrian tracking by adaptive re-identification. In: 2023 IEEE international conference on image processing (ICIP). IEEE, pp 3025–3029
11. Cao J, Pang J, Weng X, Khirodkar R, Kitani K (2023) Observation-centric sort: rethinking sort for robust multi-object tracking. In: Proceedings of the IEEE/CVF conference on computer vision and pattern recognition, pp 9686–9696
12. Du Y, Zhao Z, Song Y, Zhao Y, Su F, Gong T, Meng H (2023) Strongsort: make deepsort great again. IEEE Trans Multimed 25:8725–8737
13. Jocher G, Stoken A, Borovec J, Changyu L, Hogan A, Diaconu L, Poznanski J, Yu L, Rai P, Ferriday R (2020) Ultralytics/yolov5: v3. 0, Zenodo

14. Wang C-Y, Bochkovskiy A, Liao H-YM (2023) YOLOv7: trainable bag-of-freebies sets new state-of-the-art for real-time object detectors. In: Proceedings of the IEEECVF conference computer vision pattern recognition, pp 7464–7475

MIX
Papier aus verantwortungsvollen Quellen
Paper from responsible sources
FSC® C105338
FSC
www.fsc.org

If you have any concerns about our products,
you can contact us on
ProductSafety@springernature.com

In case Publisher is established outside the EU,
the EU authorized representative is:
Springer Nature Customer Service Center GmbH
Europaplatz 3, 69115 Heidelberg, Germany

Printed by Libri Plureos GmbH
in Hamburg, Germany